Guardians of Eden

EAST CUMBERLAND AGRICULTURAL SOCIETY.
PENRITH DISTRICT.

Premium 1. TO the Tenant and Occupier of the best
Managed Farm, above 150 acres, ... £8 0 0
2. Do.   Do. above 50 acres, ...4 0 0
3. Best effect from £10 expended in
   Draining, ...4 0 0
4. Best Crop of Turnips,......................3 0 0
5. Best General Farm, Live Stock,........5 0 0

Claims for Premiums, 1, 2, and 3, to be sent to the Secretary, on or before Tuesday the 1st of July, 1834; and Claims for Premiums 4 and 5, to be sent on or before the 1st of September, 1834.

J. BROWN, Secretary, Penrith.

CARLISLE DISTRICT OF EAST CUMBERLAND AGRICULTURAL SOCIETY.

A MEETING of the Members of the above Society will be held at the *Bush Inn*, on FRIDAY the 12th inst., at the hour of 10, A. M., to open the SECOND Subscription for the Exhibition 1835; and when other important Business will be decided upon.

At this Meeting a List will be produced of the Premiums and Sweepstakes (public and private) for the PENRITH EXHIBITION, to take place on the 26th inst.; when all Persons desirous to enter their Stock, will have it in their power to do so; and if they neglect, it will subject them to the inconvenience of communicating with the Penrith Secretary, up to the 24th inst., when the Entries finally close.

WM. PATRICKSON, Crosby, } Secretaries.
THOS. DONALD, Linstock, }
September 5, 1834.

From The Carlisle Journal

Guardians of Eden is published by Penrith Agricultural Society as a new millennium celebration of the role and influence of agriculture in East Cumbria in the 166 years since the first Penrith show or exhibition which was held on Friday, 26th September, 1834, in a field adjoining Croft House.

Compiled by Andrew Humphries and Bill Mossop

Text by Andrew Humphries

December, 2000

The publishers wish to thank all those who kindly contributed material for this book and without whose generosity the project would not have been possible. The amount of material inevitably required some difficult choices to be made on the final content.

Specific acknowledgement for use of protected material is due to Lightwork, Winton; the Institute of Rural History at the University of Reading; Dixon Printing Co. Ltd, Kendal, and the Cumberland and Westmorland Herald.

First published December, 2000

Copyright © Penrith Agricultural Society

Text copyright © Andrew Humphries

All copyright © rights asserted

ISBN 0 9518552-3-9

All rights reserved. No part of this publication may be framed and sold to a third party, or reproduced, stored in a retrieval system, or transmitted, in any form or by any means, electronic, mechanical, photocopying, recording or otherwise, without the prior permission of the publishers. Such permission, if granted, is subject to a fee depending on the nature of use.

Published by the Cumberland and Westmorland Herald.
Printed by Reed's Ltd, Penrith

CONTENTS

The millennium show	6 - 8
Development of local shows	9 -17
Fairs, droves and markets	18 - 24
The cultural landscape	25 - 33
Livestock	34 - 45
Rural life and the farming year	46 - 66
People	67 - 73
The role and contribution of women	74 - 80

Millennium show— 6

Penrith show 2000

▲ Penrith Show 2000 President Jamie Fisher and his wife, Marjorie, of Yanwath Woodhouse, Eamont Bridge, presents the Champion of Champions trophy to Jackie Marshall, Kirkpatrick Fleming, whose Clydesdale heavy horse was judged the best livestock exhibit.

KEY INGREDIENTS COME TOGETHER ran the Cumberland and Westmorland Herald headline over the report of the millennium Penrith Show — and so it was. Fine weather, an excellent crowd and superb entries surpassed all expectations in a display that would surely have amazed those pioneers of 1834.

At the centre was the livestock with champions of national standard. Yet the underlying purpose of the show goes beyond the winners. Tribute needs to be paid to those who did not win. The success of the show in the longer term is about participation by competitors and those who come to see the displays.

Town and country can and do meet at the show and it is to the credit of the committee that Penrith is still an agricultural show. Yet it is interesting and welcoming to non-farming visitors who can come closer to those who continue to produce food and other products of the highest quality. Appropriately there were links with earlier shows, not only in the livestock lines but in the farm management awards where a premium for maize is further evidence of continuing evolution in farming practice.

As in many facets of life success is largely in the preparation. The committee, under chairman Sid Ivinson, and all the competitors should take satisfaction in having planned and prepared well. This involved considerable thought and effort in putting together the excellent pageant.

The number and range of classes compared to those of the early years is remarkable and it is worth underlining the value of the entries in all categories of the show — livestock, children's, craft, industrial and horticultural offered such a width of interest to the visitor. The trade stands and demonstrations were integral to the experience which the President Jamie Fisher so clearly summarised as "a great day, a great show."

◄ Julie Wilson with the YFC champion, a 6 months Belgian Blue X

▲ Sally Dent and Laura Pepper with the Shorthorn champion, Winbrook Jill 116, shown by G. A and D. W. Dent, Winton.

▲ On a Samson model S25 Sieve Grip tractor dating from the early 1900s, (25bhp at 650rpm; 10hp at the drawbar), James Fisher leads the millennium show parade of vintage farm vehicles.

THE TROPHY WINNERS

Sheep: Suffolk champion — R. K. and M. D. Denby, Longtown; reserve — J. J. and L. Hewson, Dumfries. Bluefaced Leicester — Messrs Raine, Kirkoswald; reserve — R. C Brough, Longtown. Swaledale — Ruth Stevenson, Kaber; reserve — H. H. Harrison, Tebay. Mule — Smith and Jackson, Haltwhistle; reserve — Messrs Raine.

Herdwick and reserve — J. and D. Richardson, Lamplugh. Jacob — C. Richardson, Ulverston; reserve — J. Corrie, Clifton. Texel — R. and E. Campbell, Cockermouth; reserve — J. R. Beattie, Wigton. Rouge De L'Ouest — J. and S. Wilkinson, Richmond; reserve — J. Harrison, Catterlen. Charollais — J. J. Wales, Raughton Head; reserve — M. and A. Soulsby, Temple Sowerby. Any other breed — J. W. and K. M. Davison, Hexham; reserve — S. Hunter, Newbiggin, Stainton.

Scotch Blackfaced — H. Smith, Gisland; reserve — W. R. Twizell, Northumberland. Lleyn — Edenhall Estates, Edenhall; reserve — J. Harrison, Catterlen. Butchers' lambs — J. Teasdale, Melmerby; reserve — J. G. Miller, Penruddock. Supreme sheep — Bluefaced Leicester, Messrs Raine; reserve — Texel, R. and E. Campbell, Cockermouth. Arable champion — M. Holliday, Langwathby.

Goats: Miss M. E. Clark, Clifton; reserve — Mrs. M. G. Johnson, Penton.

Cattle: YFC Club — W. Patterson, Appleby; reserve — R. Fisher, Yanwath. Holstein — S. L. and P. S. Bell, Bowscar; reserve — C. Dent, Kirkby Thore. Shorthorn and reserve — G. A. and D. W. Dent, Winton. Jersey — J. and J. Chester, Wigton; reserve — T. R. Savage, Bolton, Appleby. Dexter — Mrs. Raybould, Bishop Auckland; reserve — Mrs. Schofield, Ivegill. Supreme dairy champion — S. L. and P. S. Bell; reserve — J. and J. Chester.

Limousin — D. B. D. and C. A. Edgar, Ousby; reserve — J. M. Thompson, Melmerby. Charolais and champion — J. Holliday, Hunsonby. Blonde D'Aquitaine — A. Hall, Darlington; reserve — Paragon Blondes, Hexham. Simmental — J. A. Tallentire, Great Asby; reserve — Messrs Addison, Kings Meaburn. Any other beef breed — C. S. Fletcher, Appleby; reserve — N. Luckett, Carlisle. Highland — D. Hodgkiss, Embleton; reserve — E. Halford, Stocksfield, Northumberland.

Prime beef — N. E. Slack, Newby; reserve — T. Hollingswood, Mold, N. Wales. Supreme beef — D. B. D. and C. A. Edgar; reserve — N. E. Slack.

Horses: Riding pony — Mrs. K. Allen, Milnthorpe; reserve — Mrs. G. Poole, Wigton. Fell ponies — Mr. and Mrs. A. W. Morland, Tebay; reserve — Mrs. P. Randell, Southwaite. Shetland Ponies — Miss A. Edge, Old Hutton; reserve — G. and D. Park, Gilcrux. Mountain and moorland — Mrs. L. Crayston, Ulverston; reserve — B. Benz, Maulds Meaburn.

Hunters — Miss M. L. Stephenson, Darlington. Show ponies — Mr. and Mrs. C. Jackson, Lancashire; reserve — Mrs. A. Higham, Silverdale. Heavy horses — Mrs. J. Marshall/Mrs. C. Halliday, Kirkpatrick Fleming; reserve — M. Robinson and Son. Private driving and marathon — Mr. Judson; reserve — Dr. M. M. Sanderson, Lazonby.

Industrial: Most points, produce — S. Fisher, Eamont Bridge. WI cup — Plumpton. Most points, handicrafts — Miss D. Irving, Skelton, and Mrs. I. Johnston, Skelton. Most points, floral art — A. Douglas, Stainton, and A. Morley, Southwaite.

Horticulture: Most points — G. A. Atkinson, Winskill. Most points, vegetables — E. J. Glendinning. Juniors: Most points, under five years — Duncan Westbrook; five to eight years — Sian Edmondson; nine to 12 years — Wayne Cartmel and Robert Teasdale; 13 to 16 years — Kelly Smith.

Millennium show — 8

◀ *Mark Phillips with the Limousin and interbreed beef champion, a bull owned by Doug Edgar, Ousby.*

Susan Fisher, Eamont Bridge, with an award-winning cake. ▼

David Raine, Old Parks, Kirkoswald, displays the overall sheep champion, a Bluefaced Leicester five-shear ewe. ▲

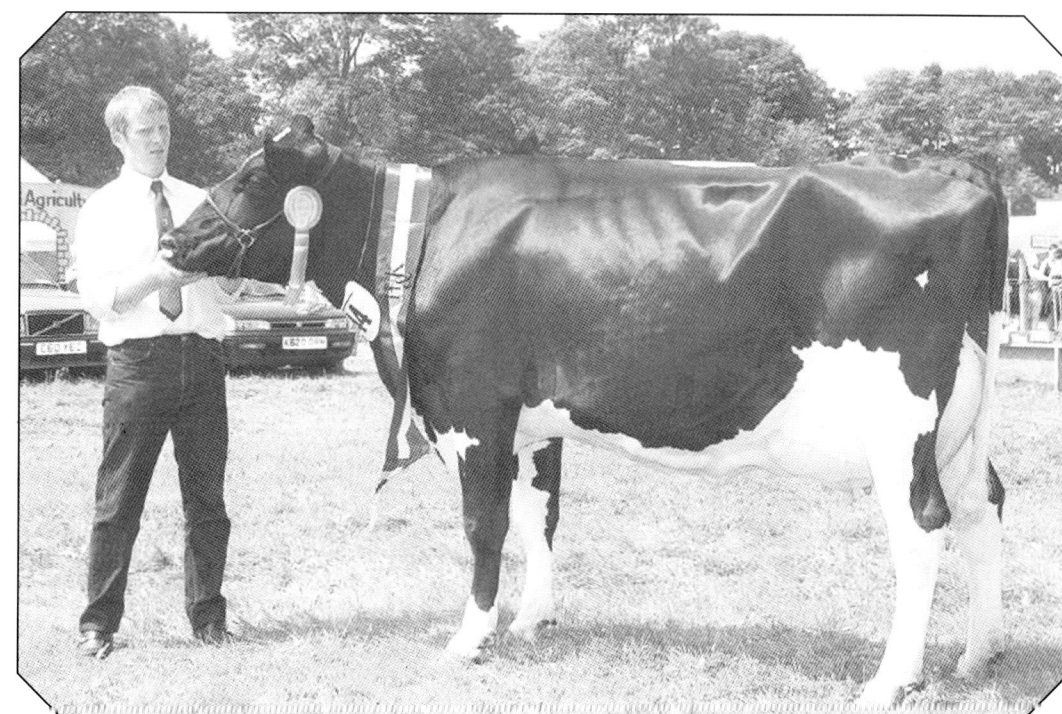

Stephen Bell, Forest Hill, Bowscar, and the Holstein interbreed dairy champion, a four-year-old cow, Holmland Delta Ruby II. ▶

Development of the local shows

▲ *Nowhere has finer settings than Eden for rural shows and events. This panorama of the Ullswater valley, taken in 1959, shows the beautiful location of the King George V Playing Field at Patterdale with the annual Ullswater Dog Day in progress.*

EARLY AGRICULTURAL societies were generally established as the means of exposing rural communities to changes in farming practices. Their rise in the late 18th and early decades of the 19th centuries coincided with growing demands for food from an urbanising society.

Almost without exception the early successes were due in large measure to the efforts of individuals of remarkable energy, exceptional enterprise and public spirit.

Cumbria's good fortune lay in the person of John Christian Curwen, of Workington Hall. With the profits arising from his coal mines he began to acquire land on a large scale. His awareness of the limitations of agricultural practice of the time arose from the cost of feeding the horses working in the collieries.

A failure in the hay harvest in 1801 and the high cost of oats led him to look for alternatives and for the next two decades he undertook a major programme of experimentation which brought both local and national recognition. The Workington Agricultural Society, established in 1805, was a national focus and attracted the leading agriculturalists from all corners of the kingdom, leading Sir John Sinclair the first president of the Board of Agriculture to refer to Curwen as a "Field Marshall in the armies of agriculture". [i]

▲ *Native pride, 1907. An award winning Herdwick ewe and gimmer lamb.*

Lord Lonsdale's prize ox, as illustrated in the Agricultural Gazette in January, 1875 — an example of how leading landowners were demonstrating the potential for improving livestock.

Among the successful competitors at Workington were a number of Eden farmers from the first show in 1806, pre-dating the Royal Show by some 30 years.

The first Penrith Show, 1834

Perhaps the stimulus to initiate a show at Penrith began with the competitive spirit among the local improvers led by Lord Lonsdale. In 1829 John Buston went to Dolphenby [ii] and brought with him two quality improved Shorthorn bulls, Crofton's Cripple and Young Rockingham. This led to local comparisons being openly expressed. The effect was a private sweepstake at the first Penrith show for a prize of 5 guineas. Shortly before the appointed day the rumour was abroad that at the time of Richmond races Lord Lonsdale had bought a new bull for 100 guineas from the elite herd of Colonel Cradock, at Gainford, near Darlington.

"When the great unknown descended from his van on to the show ground in the shape of a three-year-old scion of Thorpe and Cherry, the owners of his opponents saw that their chances were quite out." Mr Buston had sent Sir William and Priam and Wallace from the Denton and Troutbeck herds. [iii]

What better start could there have been than the attraction of seeing the elite new Shorthorns at their best.

The East Cumberland Agricultural Society was formed in 1833 [iv] with secretaries appointed to its Carlisle and Penrith districts and in the following year the society promoted the first Penrith show. On June 28th, 1834, a notice in the Carlisle Journal announced the first show. A further notice on September 6th referred specifically to the Penrith Agricultural Society, giving the first show date as Friday, September 26th, the venue being a field adjoining Croft House. [v]

The classes for livestock included Shorthorn and Galloway cattle and Leicester and Blackfaced sheep. Pigs were not specified by breed and the horses were defined as to whether for agricultural purposes or for coaching (or Cleveland type). The absence of Longhorn classes is indicative of the changes underway in the pattern of breeds in the area.

Mr Benn's magnificent bull

The reports referred to the stock as superior to anything previously seen in the county. William Blamire MP, a noted improver, declared that with the exception of the Highland Society's shows and those at Smithfield, the first Penrith Show was inferior to none in the Kingdom. He added that he did not think until that show day that such a bull existed in England as that shown by Mr Benn, the Lowther steward. The sheep were also considered to be excellent including the Leicesters with "coats sleek and glossy as silk".

In addition to the formal classes there were a number of sweepstakes for each of which the stake was required prior to the show. The classes for sweepstakes fell into those instituted by the society

▲ *The pavilion for the Royal Show held at Carlisle in 1902 — last of the travelling shows.*

and private sweepstakes. The latter included competitions for Herefordshire cattle, an indication of a growing interest in specialised beef breeds. The principle of the sweepstake is that the stakes are pooled by the competitors and the winner takes all, in effect "sweeping everything away".

For many of the early societies an important element in the prize list was the recognition of progressive farming techniques which could not easily be incorporated into show day. Classes at the first Penrith show included farm management premiums for large and small farms and specific prizes for drainage, turnip production and general stock of store cattle. The last mentioned was awarded to John Buston, of Dolphenby, who was toasted at the dinner as the individual who "seemed to have run away with the best part of the prizes" (if not the sweepstakes). Whilst local farmers took the bulk of awards, competitors came from places including Abbey Holme and the Netherby estate.

A section of prizes under the heading of rewards gives a direct insight to rural life at the time. Twelve competitors entered for the premiums offered for service in one situation for men, which was awarded to Thomas Byers whose 61-and-a-half years with Matthew Atkinson at Temple Sowerby eclipsed the competition. Sarah Dean was awarded the premium for women, for 35 years with William Noble, of Burrowgate.

▲ *Penrith Show in 1964 on a field in Bridge Lane which is now the site of Penrith New Hospital.*

▲ *An impressive line-up of cattle at Penrith Show on the Foundry Field in the 1950s.*

▲ G. W. Dent, Kaberfold, Kirkby Stephen, one of Eden's most famous breeders of Shorthorn Cattle.

▲ King Edward VIII among his people at Thirlmere in 1936.

At this period the cost of supporting the poor was a concern in all parishes and a premium was offered as an encouragement for the labourer in husbandry who had brought up the greatest number of legitimate children without parochial relief. The £3 prize went to Alexander Armstrong who had raised 13 children all living, and had lived 33 years with Mr George Syme, of Redkirk. This class attracted no fewer than 10 entries. [v]

Differing views on the ideal animal

From the beginning, the world of shows was permeated by different views on the ideal animal and the principles of judging. Fatstock particularly provided a real challenge to judges. At the first show the Carlisle Journal noted that "in all points denoting excellence, it would be utterly hopeless to look for anything like universal satisfaction; it is but justice, however, to say that the judgements given by these gentlemen were all but universally satisfactory". Some things never change!

The attendance on a day on which "it began to rain at an early hour in the morning and continued with little interruption during the whole day" was officially 526 besides the officials. The dinner advertised at the George Inn for three o'clock was held back until five, caused by the numbers of stock to be judged. About 160 gentlemen sat down, with Henry Howard, the High Sheriff of Cumberland, in the chair. Lord Lonsdale having attended the judging but being too infirm to stay for the dinner, sent his apologies and good wishes. The chairman's speech was one of encouragement to meet the difficulties besetting agriculture in a positive vein, in the assurance that increasing food production was a noble and fitting aim for the industry. Well in excess of 20 toasts were given and many songs enjoyed before the meeting concluded at about eleven o'clock. [vii]

The pages of the local press provide insights into a changing industry striving to improve production and productivity, responding to the challenges of imports and the incentives to emigrate.

Carlisle Journal, January 11th, 1834

While the British Landowners are asleep, the Americans are pouring their wheat into Canada to be ground there and sent to this country as Colonial Flour.

Carlisle Journal, August 16th, 1834

The Holme Agricultural Society in West Cumberland were discussing the merits of a locally invented reaper which was commanding national attention. The machine, invented in 1832, was the brainchild of Mr Mann, of Raby Cote, the ancestor of Mrs Jamie Fisher, of Woodhouse.

Carlisle Journal, January 10th, 1835

An advertisement brought to the attention of farmers the service of a "Powerful New Steamship", the Newcastle, which would carry livestock direct from Port Carlisle to Liverpool at reduced rates with an additional 5% reduction for cash at the time of sailing. Rates included bulls 20s, fat cattle 8-12s, store cattle on deck 6s, fat sheep 1s, store sheep and lambs 8d.

Carlisle Journal, March 7th, 1835

The British American Land Company incorporated by Royal Charter have FOR SALE ONE MILLION ACRES OF LAND in farms of 100 acres and upwards. Prices from 4s to 10s per acre payable one fifth cash down on the higher priced lots, one fourth on the lower priced ... the balance in six annual instalments ...

Whilst Longhorns never made it into the premiums at Penrith, Herdwicks appeared from the second show. The show has enjoyed several venues including the Foundry Field and Football Field, moving to Winters Park about 1960, thence to Bridge Lane where the new hospital now stands and more recently to Brougham, perhaps the most attractive of settings with the amenity of the riverside, adjacent deer park and Brougham Hall. From 1846 the Society became known as the Cumberland and Westmorland Agricultural Society holding its shows at Carlisle (1846), Appleby (1848, 1852), Wigton, Kirkby Stephen (1862), Keswick, Shap (1875), Milnthorpe (1878), and Penrith (1849, 1853 and 1864). At Penrith in 1849 a prize for butter in firkins was offered and eight poultry classes introduced, including Muffled Cochin China, Wattled Cochin China, Silver Pencilled and Game fowls. At the 1852 show Lord Lonsdale gave a prize for the best essay on Westmorland Agriculture which was awarded to John Bell, barrister at law, Recorder of Appleby.

▲ *Judging in progress at Tan Hill Swaledale sheep show, high on the Cumberland-Yorkshire border in the late 1950s.*

In more practical vein the following year Wm Atkinson, of Winderwath, won the £10 prize for the best muck midden. [viii] Galloway classes were discontinued in 1860 and two years later a horse-shoeing competition was introduced. The last of the travelling Cumberland and Westmorland shows was in 1878 at Milnthorpe [ix] when the horse classes covered hunters, harness, road or field and agricultural horses. Since then the Penrith shows have all been under the auspices of Penrith Agricultural Society.

Origins of locals shows

Other shows which started in the area include Brough with its traditional fair underpinning its claim to a farming show. By 1900 the entries totalled 829 and the show continues to thrive with a particular feature today being the drystone walling competitions. Stainmore show, which ran from 1895 until 1909, attracted 345 entries to its Mouthlock site in 1906. Tup shows came and went at Milburn and Bampton which incorporated the Bampton Sheep Association for the Improvement of Mountain Sheep. Orton, founded in 1860, included classes for Whitefaced Sheep, probably the Limestone Crag Sheep, now extinct. Shap began the following year with Fell Ponies as a noted feature. [x] Joseph Taylor, recalling earlier days in 1934, described how the Brown family of the Bell and Bullock in King Street saw their ponies run away at the Kirkoswald show. A Lazonby horse dealer called Miller ran into the ring, seized the first pony by the neck and "buttocked it". [xi]

Many other communities developed local events with varying success. Shows like Skelton have achieved great success on a large scale. Dufton, founded in 1863, was at first principally a ram show and from 1876 had cattle classes. By 1909 there were over 600 entries. Known from early times as the "Fellside Royal", it stands as an example of a smaller show which has survived and prospered. Alston Agricultural Society celebrated the millennium with the 150th show of their 162 year history

The Penrith Jubilee Show of 1892

The 50th anniversary of Penrith Show was marked by the presidency of the Earl of Bective, himself an ardent admirer of Shorthorn cattle for which Penrith can claim to have been the national centre.

Eleven years earlier he had purchased three Shorthorn females at the famous New York Mills sale in 1873 for over £12,000. The 10th Duchess of Geneva alone cost some £7,000. Ten years on from the 1884 show the herd dispersal averaged only £46 1s 8d, reflecting the severity of the depression. [xiii]

At the 1884 show the report of the judges of the green crops is of interest. "One exhibitor illustrated the rapid growth of his turnips by saying that if they grew as rapidly for the next fortnight as they had in the last they would knock the walls down ..." The Shorthorn winners included the well known herds of R. Thompson, Inglewood, J Harris, Calthwaite Hall; and J. C. Toppin, Musgrave Hall Skelton. [xiv]

Ten years on in 1894, the president, H. C. Howard, devoted his speech at the show luncheon to an exposition on agricultural education. Harry Howard was fully committed to such matters and by 1896 Newton Rigg was operational. At the 1894 show there were 50 entries for the horse shoeing competition, six anvils being kept going. The Shorthorn classes were dominated by other local herds including those of W. Graham, Eden Grove, E. Ecroyd, Low House, Armathwaite and J. Handley, Greenhead.

The 1913 was to prove to be the last Penrith show before the war. A new departure was a forestry exhibition when prizes for essays were distributed by Mr E. W. Hasell, Dalemain.

The 1924 show was held on Tuesday "The Glorious Twelfth". Despite this Lord Lonsdale was present with a party. [xv] During this period many of the stock shown at Penrith were simultaneously taking national honours and confirming the quality of Cumbrian stock. Mr J. Harris's bull Oxford Duke of Calthwaite 100th won three Royal shows and made a significant contribution to the breed. Robert Thompson's Inglewood herd was visited by the Royal Association of British Dairy Farmers in 1924 and the sight of the conveyances carrying some 200 admirers from Penrith to Great Salkeld must have been a talking point in the district for many years. [xvi]

His cow Molly Millicent was widely regarded as the finest Shorthorn cow of her era, winning five championships and 33 other Royal show prizes. It is interesting to note the inclusion of Friesian Cattle at the 1924 Penrith Show, though they did not feature in the centenary show of 1935. [xvii]

▲ *The 50th anniversary of Penrith Show (as reckoned at the time) was celebrated in 1892. Pictured at that jubilee event, are (back, left to right): George Bowstead, Beck Bank, Great Salkeld; John Westgarth, Brougham; Tom Heskett, Plumpton Hall; Phillip Sowerby, Bank Hall; John Armstrong FRCVS, Penrith; and Tom Robinson, Edenhall (all stewards). Second row: T. Hills, Vale of Eden Dairy, Culgaith (judge); A. Dobinson, Williamsgill and Henry Winter (stewards); J. H. Stokes, Market Harborough (judge); W. H. Thwaytes, Woodhouse; A. S. Slack, Skirwith Hall; D. S. Ingledew, Sewborwens; Arthur Graham, The Limes; and John Davidson, Shepherds Hill (stewards). Third row: John S. Stow, Darlington; H. Moor, Burn Butts, Cranswick; E. Hall, Barton; — Campbell, Kirkudbright; Matthew Ridley, Peel Well; T. Dodds, Wakefield; W. Vickers, Howe John, Darlington, — Richardson, Smeathwaite (all judges). Front: Thomas Hudson, Whale Moor (Steward); John Thornborrow (secretary); T. Kerr, Kirkudbright (judge); and W. R. Mounsey, Penrith (steward).*

During the middle years of the show's life much of its success was due to the guidance of John Thornborrow during his secretaryship. Such was the nature of his contribution and the appreciation of members that he was made the only honorary life member at that time. It was appropriate that the Centenary show should be under the presidency of the Earl of Lonsdale. It was his forebear who had not only brought Shorthorns to the district in 1810 but who had given a lead in matters of agricultural improvement. [xiii]

The Centenary Show — cavalcade of a century

Blessed by ideal weather and record entries, the centenary show was judged to have been a memorable event for an industry facing long term difficulties. In his appeal for support prior to the event the chairman, William Bainbridge, referred to a dilemma which has challenged show committees throughout its history.

"These are difficult days for agricultural societies and it is thought by many people that a show cannot succeed without a lot of additional attractions. I hope patrons of the show agree we cater for the public, but we do not forget the primary object of the society — to promote the good of agriculture."

The success of any show depends on the preparations carried out throughout the year by many seen and unseen workers and in the support given by those holding senior office. In the centenary year it is fitting that the presidency lay with Lord Lonsdale, a consistent and generous supporter of the event. With him came the guests whom he was entertaining for the grouse shooting. The twelfth had fallen on a Sunday and the party had engaged in shooting only on the Monday afternoon on the Shap fells where the birds had been plentiful and healthy. On the Tuesday Penrith Show took priority for the group which included the ex king of Greece, the Earl and Countess of Marr, Sir William and Lady Noreen Bass, Lord Hamilton of Dalzell and the Compte and Comptesse of Preux.

A particular success was the display of non pedigree dairy cattle, giving ample opportunity for the best of the local commercial farmers to impress. The newly calved Shorthorn cow from Mr J. Smith, of Far Close, Dufton, attracted many compliments as did a stylish heifer from J. T. Robson, of Sewborwens, who also won the class for a group comprising a beast, a horse and two sheep. Official award recipients for the Dairy Shorthorn Association included Messrs Jackson, of The Wreay, for their Shorthorn cow Wreay Wildeyes whilst the Westmorland flag was carried by Messrs. H. Holme and Son for their heifer Thrimby Lady Elma.

The judge for the Clydesdales travelling from Orkney made a forced landing at Aberdeen and did not arrive until the Tuesday afternoon. Obligingly he purchased for a three figure price the first prize colt foal from Messrs H. Dobinson and Son, Leeming Farm.

Among the sheep was a quality line up of Herdwicks including successful entries from Mrs Heelis (Beatrix Potter) and R. M. Wilson, of Glencoyne, Ullswater. Swaledales were represented locally by the winners of the group class, J. S. Armstrong, of Outhwaite, Renwick, with the leader of the Rough Fells from the stock of Sir S. Scott, Yews, Windermere. Of the breeds used to produce half breds representation came from the Border Leicester and Wensleydale flockmasters, but the Bluefaced Leicester era was not yet underway.

Shows — 15

▲ Lord Lonsdale's party at the centenary Penrith Show in 1934. The former King of Greece is seated on the bottom row, sixth from the left, with Lady Lonsdale (looking away from the camera) on his left.

"Classical company"

The report in the Cumberland and Westmorland Herald notes the impressive display from John Dargue, of Burneside Hall, following earlier successes at the Royal, Great Yorkshire and Royal Lancashire shows. He brought his team to the show and they were much admired. Local breeders H. Rumney, of Pallet Hill, and W. L. Parkin, of Sleagill, had to be content with third prizes in such "classical company".

Cumberland pigs attracted 25 entries. Messrs Gardhouse, of Wigton, took the major honours with local success for Messrs Masheter, Romanway, Plumpton, and Mrs Carleton-Cowper, of Eamont Bridge. In the poultry section entries came from as far as Huddersfield and Matlock.

Strong support was evident for the butter and breadmaking with 21 and 38 entries respectively. Particularly busy was the Newton Rigg tent with a display on warble fly control and bacon pig carcase comparisons from Cavaghan and Gray.

The poultry classes were held in Tinklers premises with the top award, Lady Mabel Howard's challenge cup, going to a tasselled grey Old English Game cock in faultless condition exhibited by Greenhow and Hartley, of Annan. The horse leaping attracted 18 entries. [xix]

---

i Personal communication from Michael Curwen.

ii The Dolphenby referred to is at Coltham Stob in County Durham. This can be confirmd by reference to the 1810 Ketton Sale where Mr Buston and the Earl of Lonsdale were in competition for the stock on offer. Both were successful and the event may have encouraged them to wager in a sweepstake to settle their rivalry. In 1837 Mr Buston moved with his herd to Wharton Hall, Kirkby Stephen.

iii H. H. Dixon, Saddle and Sirloin, London, 1870, p 91.

iv Carlisle Journal, Aug 1st, 1835.

v Carlisle Journal, Sept 6th, 1834.

vi Ibid

vii Ibid

viii Cumberland and Westmorland Herald, Aug 11th, 1934

ix Bingham R., from Fell and Field, Kendal, 1999, 63.

x Garnett F. Westmorland Agriculture, Kendal, 1912, p 17.

xi Cumberland and Westmorland Herald, Aug 11th, 1934.

xii Ibid.

xiii Garnett F. op cit. pp 86, 87.

xiv Cumberland and Westmorland Herald, Aug 11th, 1934.

xv Cumberland and Westmorland Herald, Aug 18th, 1934.

xvi Article by Miss J. Stubbs in the Journal of the Royal Association of British Dairy Farmers, 1924, pp 67-79.

xvii Penrith Show Catalogue, 1935.

xviii Tribute by William Bainbridge in Cumberland and Westmorland Herald, Aug 11th, 1934

xix Cumberland and Westmorland Herald, Aug 18th, 1934.

▲   *A shot taken 40 years ago of Dufton Show — "The Fellside Royal" — in its picturesque setting at the foot of the Pennines.*

▲   *An attractive photograph of busy Hesket-new-Market show in 1959.*

▲ *Officials and competitors pictured at Ousby Sheepdog Trials in 1959. The secretary of the event, T. M. Wales, won the local trials and M. Cook, Kildale, Whitby, the open class.*

▲ *Onlookers engrossed by fellow competitors working their sheep at Alston Sheepdog Trials in 1955.*

Fairs, droves and markets

▲ *Cattle crossing the Solway Firth, from a painting by Sam Bough RSA.*

THOMAS GRAY in his Journal of a Tour wrote on the 30th of September, 1769: "A mile and a half from Brough on a hill lay a great army encamped. It was the Brough cattle fair. On nearer approach appeared myriads of horses and cattle on the road itself; and in all the fields round me a brisk stream hurrying across the way; thousands of clean healthy people in their best parti-coloured apparel, hastening up from the hills and down the fell in every side glittering in the sun and pressing forward to join the throng." [i]

Such observations are a reminder of the place of fairs and markets in the lives of the people of Eden who made them and were in turn shaped by them. Generally fairs were annual and markets weekly though frequently granted at the same time. Enquiry suggests that many of the fairs sited on hilltops such as at Rosley, Brough and Appleby may mark sites of significance to earlier inhabitants.

In the 12th and 13th centuries large numbers of Charters were granted. These did not necessarily mark the establishment of the fair but rather a formalising of permission. Such arrangements ensured payment of tolls in return for the right to hold the fair. The conditions attached to such events included the appointment of officers. These might include Swine and Goat Lookers, Searchers of Weights and Measures and Lookers of Flesh. In May 1855 the Health Authority seized a diseased carcass from a butcher at Orton. The meat was burned in the market place and the butcher fined £5. At Appleby a jury would be called at the nearest public house and the matter dealt with there and then. [ii]

Kirkby Stephen's ancient Luke Fair

Today few reminders remain in a physical sense of arrangements, which shaped the economy of many of our market towns and villages. Yet some can be identified. For example the width of the main street in Kirkby Stephen seems to act as a reminder of the sheep pens on either side for the October Luke Fair.

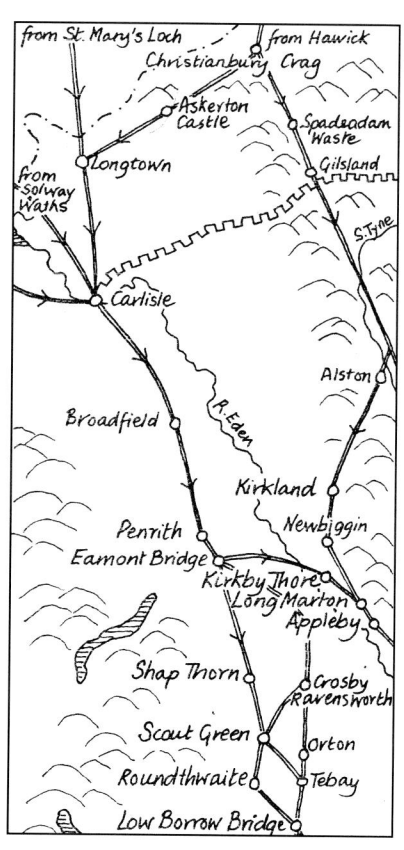

▲ *The old cattle droving routes.*

The importance of the trade in livestock is difficult to appreciate. The Brough Hill gathering was probably the largest fair for the sale of Fell and Dales ponies which during the 19th century were critical to the needs of the Durham pit owners. The two day cattle fair at the end of October attracted up to ten thousand head. Such events were also a rallying place for the north country gypsies. No less significance applies to Appleby with its market charter granted by King John. In 1750 a fair was instituted under the seal of the Mayor for "Horses, Sheep, also Black Cattle if it so pleases God to cease the Distemper which now rages among them" on the first and second days of June thus establishing the "New Fair" on Gallows Hill. [iii]

The fairs at Shap and Mardale

Shap too had a fair held in May with a toll of 1d per beast, which by ancient custom was paid as a perquisite to the head shepherd of Shap Abbey. [iv] Nearby a fair was established at Mardale at the head of Haweswater in 1824 and was followed by sports. Undoubtedly there was competition for the trade between the various events. In 1829 a writer notes that at the Brough Hill Fair on Oct 2nd "several of the spice wives move away in the afternoon to Kirkby Stephen where there is a very large fair better suited for their trade". [v]

Over time the fairs declined but until the opening of the railways they flourished where they lay on the great droving routes. Inevitably the growth in population and the development of urban and industrial centres challenged the farming population to provide for them. From 1707 the union of the crowns gave the green light to droving as a more legitimate and often profitable alternative to reiving. [vi] Opportunity came and was grasped by many Cumbrians. The art and mystery of the drover was respected since laws dating from Elizabethan times controlled them. They were required to be thirty years of age, a householder and to wear a uniform. Understandably considering their responsibilities they were exempt from the disarming acts. Great skill was needed to negotiate routes in a countryside whose shape was changing in response to enclosure. Balancing the needs of stock with the need to arrive in time for the sale perhaps 150 miles away was a skilled occupation.

Animal welfare legislation provided for legal redress in cases of cruelty to animals. Negotiating prices, undertaking the droving and making payments of grassmail for overnight rest on the stances were attributes found in many Cumbrian farming families.

The droves began in the furthest reaches of the British Isles including the Hebrides, Ireland and Galloway. Like the tributaries of a great river the ever-growing numbers flowed from the remote uplands to the industrial towns of Yorkshire and Lancashire, the Midlands and ultimately the metropolis.

▲ *The hustle and bustle that was once Brough Hill Fair is captured in this painting by Donald Wood, of Warcop.*

Inevitably the coming of the railways changed everything in the space of a few years. At a Highland Show dinner in the mid 19th century the toast was offered "steam is your highland drover". [vii] Few signs remain. In Penrith is Drovers Lane near the head of which is the Grey Bull Inn. Past its door would have come thousands of Southbound drovers and their charges, bringing stock from the Falkirk Tryst, from Dumbarton and Doune, from Rosley and Carlisle and on to Brough or Garstang and the South.

Ready Cumberland money at Falkirk

The Cumbrian writer on 19th century agrarian life H. H. Dixon recounted a buyer at Falkirk, in all probability a Cumberland drover, who purchased 20 score of middle horned cattle on the muir. He did not wait to be asked for the money. Off came his coat and waistcoat, and out came his penknife. In seconds he had unstitched the waistcoat lining and taken the amount of £4,400 in Bank of England notes from within.

▲ *The 560th Luke Fair at Kirkby Stephen, October 27th, 1911.*

Fairs and markets — 20

Busy Troutbeck auction mart in its heyday 40 years ago, alongside the Penrith-Keswick railway line. Both the mart and line are now closed. Inset: Presentation at Troutbeck to long service Penrith Farmers' and Kidd's auctioneer Jack Thorburn on his retirement.

Many of the best North of Scotland Cheviot wethers offered at the September tryst came to Cumberland. Among the leading buyers at Falkirk Dixon notes Young and Butterfield of Penrith. Nelson and Boustead dealt entirely with sheep. Nelson held shows at the Bull's Head Plumpton where Cumberland and Westmorland famers gathered to meet them. [vii]

For droves passing through, the cost of overnight grazing adjacent to an inn was an important consideration. Broadfield and Edenhall were such places. An account book of Aug. 18th, 1712, contains the entry:

   "then owing and indebted by me Mr William Jonston for the gras of 260 drove cattle unto Sir Chris. Musgrave the sume of one pound Eight Shillings which sume I promise to pay upon demand as witness my hand

   John Nelson in his mark.

(Source: An account book of Sir Christopher Musgrave of Edenhall, bound in old vellum, private collection)

The following letter was sent from Lord Hothfield to his Agent at Appleby on Oct. 24th, 1817.

   "Sir,

   The cattle arrived yesterday. Only four or five appear a little tender in the feet. One of the heifers calved, I understand, early in the journey, but she is come with the others. They have had fare weather, all the way excepting the last 2 or 3 days.

   The whole are in very good condition and they appear to me, particularly the bullocks, as good as I have ever had. I have had larger but I think the size of these is preferable in proportion. I certainly never had any arrive in such good condition. My Bailiff thinks the Heiffers are also very good, but I have hardly had an opportunity of looking at them."

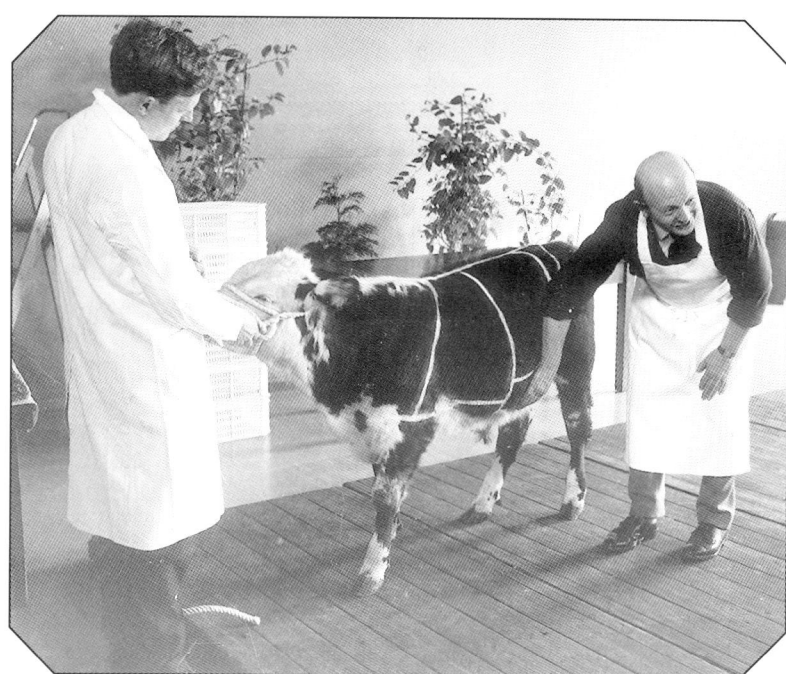

▲ *Penrith butcher Joe Clark demonstrates the cuts of a beef animal at Newton Rigg c 1970.*

Fairs and markets — 21

▲ *Two photographs from the late 1950s when transport of stock by rail was still common. Horses and a bull are loaded at Penrith goods yard.*

One of the traditional drovers was John Dodd born at Skirwith in 1799 who was a regular buyer of sheep at the Falkirk Tryst on Stenhousmuir which was held yearly on the second week of September. In the year 1858 he bought 1800 sheep and left Stenhousmuir on the 13th having paid tolls of £1.3s.3d. Further tolls were paid at Addiewell, West Linton and Tushielaw. On the 22nd he paid tolls at North House and at Newcastleton on the 25th. The total cost of tolls came to £11.15s.3d.

The drove ended at Askerton Castle in Bewcastle where the shepherds of the syndicate for whom he was buying met them. Seven drovers were paid some at £2.16s.0d and others at £2.9s.0d each. Those with dogs received an extra 6d per day. The total cost of the drovers came to £21. There would have been costs associated with accommodation for the drovers at inns en route.

Grazing costs at the overnight stances came to £37.6s.6d, bringing the total to £70.11s.9d or 9d per sheep. Travelling allowed for a rest twice daily and a complete rest every third day. The sheep were blackfaced wethers aged four to five years and were fattened on turnips. They cost 25s - 26s each and were sold the following spring for 40s - 42s each. [ix]

Naturally over time markets replaced fairs, and from the mid 19th century auction marts replaced both for livestock selling. Just as the arrival of the railways spelled the demise of droving, so it created new opportunities for farmers.

Mardale Green butter for the city

For agriculture in Cumberland and Westmorland the mid 19th century rail network provided direct links to centres of population and consumers. The possibility of local finishing on a wider scale and adding value became a real option. Up until this time milk had played only a small part in the farming economy nationally and was relatively unimportant in Eden. By 1855 some 3,000 lb. of butter was sent weekly in summer by rail from Mardale Green to Manchester.

A carrier picked up baskets from the scattered farms for delivery to the station. By the mid 1880s the competition was so keen among the middlemen that other weekly markets were established at places which included Ravenstonedale, Shap Orton and Warcop [x]

The first butter market at Orton was held on August 2nd 1864 when between 2000 and 3000 lb. were shown. In the 1860s farmers' wives in Westmorland taking loads of poultry, butter and eggs to sell in nearby markets were often intercepted by dealers from Manchester and other manufacturing towns and would dispose of the entire content of their cart in roadside negotiations. [xi]

◄ *After being rounded up on the local fells, broken and unbroken ponies are driven loose headed through the streets to the annual Cowper Day sales at Kirkby Stephen. In the 1960s up to 400 Dales cobs and ponies were entered for sale.*

▲ J. G. Bellas, Bedlands Gate, Newby, and sons Eddie and Geoff proudly display two Shorthorns at a Penrith sale in the early 1960s.

▲ Auctioneer Norman Little (stick in right hand) casts his eye over the stock at Penrith mart, in company with buyers and onlookers.

▲ *The sale of George W. Dent's Northern Dairy Shorthorn herd attracted a crowd of 500 to Wharton Hall, Kirkby Stephen, in October, 1965. Top price was 325gs for Brook Pamela 42nd.*

For those Eden farms with a good water supply the option to sell liquid milk was even more attractive. Newcastle provided the first opportunity at a cost of 1d per gallon up to 100 miles. By 1894 Mr Thompson of Hartley Castle had organised a group to send milk daily to Liverpool and London. [xii]

For livestock the emergence of auction marts represented a major step into organised selling. For the buyer they provided assured supplies of stock and for farmers a fairer pricing mechanism and cash at the point of sale.

A stimulus to auctions came in 1845 with the repeal of the tax on movables sold by auction. Prior to this only on farm sales were exempt. This change in the law combined with the development of the rail network encouraged the establishment of marts adjacent to railways.

One of the first in Britain was the covered mart at Canny Croft Penrith established by Mr Thornborrow. In 1849 Richard Harrison married Ann Moorhouse widow of Newton Rigg and in the following year purchased the Penrith mart. In 1870 he erected the Agricultural Hall in Castlegate [xiii] and subsequently sold out and moved his business to Carlisle, where he began the Botchergate Mart the forerunner of Harrison and Hetherington. In 1987 Penrith Farmers and Kidd's moved to their new premises near Junction 40 of the M6.

A national institution

Other mart premises followed including Lazonby and Troutbeck. The "grey lamb" sales at Lazonby are a national institution presenting exceptional shows of breeding and feeding lambs in the autumn.

Troutbeck until its closure was an important economic and social gathering point for the shepherds and sheep of the Lake District. Kirkby Stephen is synonymous with Swaledale breeding sheep and can claim to be the main centre for the breed. Further north Southwaite auction adjacent to the main line railway provided an outlet for Cumberland stock and attracted feeding lambs for its special sales from Dumfriesshire, Peebles and Roxburgh.

Undoubtedly dead-weight selling and direct marketing have diverted trade away from the auctions which themselves have diversified with success. They too are engaged in agricultural adjustment alongside their farming clients.

▲ *Penrith mart foreman Dick Turner chats with RSPCA Inspector Brown, 1954.*

---

i Eds. Toynbee and Whibley, Gray Thomas Correspondence Oxford 1935.
ii Garnett F Westmorland Agriculture 1800-1900 Kendal 1912 pp 132,133.
iii Ibid. p 114.
iv Ibid p 130
v Ibid p 121.
vi Bonser K J The Drovers, London, 1978 pp75, 124.
vii H H Dixon Field and Fern Scotland South, London 1865 p 36.
viii Ibid. p 303.
ix Bonser op cit. pp 86, 87.
x Webster C, The Farming of Westmorland, Journal of the Royal Agricultural Society of England, 2nd Series Vol 4 , p 13.
xi Horn P, Victorian Countrywomen, Oxford, 1991, p 109.
xii Humphries A B Agrarian Change in East Cumberland 1750-1900 unpublished M Phil Thesis University of Lancaster 1993.
xiii Kelly's Directory 1921.

Fairs and markets — 24

Southwaite Auction Mart.

Adjoining the L. & N.-W. Ry. Station ; 7 miles from Carlisle and 10 from Penrith.

ANNUAL SIXTH
SPECIAL SALE
OF ABOUT

371 Shorthorn and Polled Cattle, and 266 Sheep,

On Friday, 6th July, 1906,

Commencing at **10-30** o'clock with Harness, followed by Pigs, Sheep, and Cattle.

All the Stock will be sold through one ring, and further consignments for this Sale are solicited.

IMPORTANT NOTICE.

It is particularly requested that Vendors or their Servants will have their Stock drawn for Sale on leaving the Pens, and they must assist in getting them to and from the Sale Ring.

They must also remain in attendance until the lots are delivered, as all Stock is at the Seller's risk until it is taken delivery of by the Buyer. In cases where it is required that Stock should be trucked Sellers must assist in so doing.

The Sale is for Cash, and no lot shall be removed off the premises without a pass from the Cashier.

JOHN THORNBORROW & CO.,
AUCTIONEERS.

Head Offices : St. Andrew's Churchyard, Penrith.
Branch : 16, Main Street, Keswick, and at
Red Lion Hotel, Carlisle, on Saturdays.

Telegraphic Address : Thornborrows, Penrith.
Nat. Tel. : 095.

Luncheon will be provided at a reasonable charge by Mr. Graham, at the Station Hotel.

TRAIN SERVICES.

A Train leaves Carlisle at 9-55 a.m., returning from Southwaite 2-25 p.m. and 4-35 p.m. Trains leave Penrith at 9-55 a.m. and 11-22 a.m., returning from Southwaite 2-45 p.m. and 5-11 p.m.

There is a Postal Telegraph Office at Southwaite Station.

At the conclusion of the Sale of Live Stock there will be Sold, under the usual Two Months' Credit Conditions, about

5 Acres of Meadow Hay Grass,

A good Crop, the property of Mr. WM. NELSON, Sceugh Hill, Southwaite, growing in Two Fields on the Sceugh Hill Farm, about 1 mile from Southwaite Station. Also

3 Acres of Old Land Hay Grass,

A very heavy Crop, growing in a Field adjoining the main road at Raughton Gill, the property of Mr. WM. WHITFIELD, Howfield.

Intending Purchasers please view before the Sale.

AUTUMN SALE FIXTURES.

AUGUST.
Friday, 31st—Annual First Sale of Top Half and Threequarter-bred Lambs at Southwaite Auction Mart.

SEPTEMBER.
Monday, 10th—Mr. E. Irving, Shap Abbey—Annual Sale of Sheep and Lambs.
Thursday, 13th—Annual Second Sale of all Classes of Lambs at Southwaite Auction Mart.
Monday, 17th—Annual First Sale of Sheep and Lambs in Holly House Garth, Pooley Bridge.
Wednesday, 26th—Prize Show and Sale of Two-year-old Colts and Fillies at Southwaite Auction Mart.
Friday, 28th—Annual Third Sale of Lambs, and Prize Show and Sale of Greyfaced Ewes and Gimmer Shearlings and Border-Leicester Rams at Southwaite Auction Mart.

OCTOBER.
Wednesday, 10th—Store Cattle, Sheep and Lambs at Southwaite Auction Mart.
Friday, 26th—Store Cattle, Sheep, and Lambs at Southwaite Auction Mart.

NOVEMBER.
Friday, 9th—Store Cattle, Sheep, and Lambs at Southwaite Auction Mart.

DECEMBER.
Saturday, 8th—Christmas Prize Show and Sale of Fat Stock at Southwaite Auction Mart.

1906 sale notices from the now defunct Southwaite Auction Mart.

▲   *The retirement stock sale in October, 1978, of Mr. John Hayton, Monhouse Hill, Welton.*

The cultural landscape

▲ *An aerial shot of the north side of Penrith prior to the building of the M6 motorway. The Greystoke road is in the middle foreground, Newton Rigg College in the centre and the main East coast railway and the A6 beyond.*

EDEN IS NOT ONE LANDSCAPE but many, each perceived differently but somehow coming together in fundamental harmony. Some 20 years after the first Penrith show the American writer Nathaniel Hawthorne visited the two counties and noted in his diary in July of 1855: "On the rudest surface of English earth there is seen the effect of centuries of civilisation, so that you do not quite get at naked nature anywhere."

Later in the month he notes: "Nowhere looks as beautiful as England, this part of England at least on a fine summer morning". [i]

Many 19th century observers recognised the rare inspirational beauty of Eden and other parts of Cumbria and the link with human activity. Even at that time people were drawn to purchase property here. "In the neighbourhood of the Lakes a new class of competitor for the ownership of the soil has arisen in the merchant princes of the manufacturing districts, who eagerly buy up any nook where they may escape from their own smoke and enjoy fresh air and bracing breezes with shooting and fishing." [ii]

In the Eden valley itself the enclosures of the early 19th century created a land market which attracted those who had made money from industry and commerce to create modest agricultural estates. Those at Brackenburgh and Low House Armathwaite are good examples.

Underlying the landscape are the natural geological features which together with the climate provide the resources which are shaped by and shape the farming community in response to political policies. We may wonder whether modern society interprets the rural landscape in this way. In the autumn of 1998 the press reported widely that the Lake District was being proposed as a world heritage site as a cultural landscape. [iii] Within only a few weeks the desperate reports from the sheep sales put that proposal into context. It seems that the cultural landscape may be admired but not understood.

The landscape of Eden is more than a particular combination of landform and fields within a framework of hedges, walls and vernacular buildings. Such a backcloth is no more than a stage set in a theatre — interesting but which only springs to life and expresses its relevance once the players have appeared and the dynamic interactions between them and their environment begun. Farming and its landscape express the interaction between economic, environmental, social and cultural factors producing unique effects to the interested observer.

Creation of our landscapes

Landscapes are seen differently by individuals and to define them is elusive. However intangible they may seem, they clearly do not simply happen; they are created and stewarded by people. Such concepts seem hard to find in the politics of an increasingly global food market and the policies of the World trade organisation. Yet at a European level the concept of Environmental and Cultural Landscape are within Agenda 2000. [iv]

Historical links between dramatic changes in farming and the effects on landscape surround us in Eden. In the Napoleonic wars of the late 18th and early 19th centuries the demand for grain stimulated massive enclosures. In North Westmorland the beautiful landscapes around Newby, Morland, Hoff, Bleatarn and Sleagill were largely shaped in a period long referred to in the area as the "Bonnyparte Time". [v]

Further North the Inglewood enclosure under an award of 1819 created the largest enclosure in the history of England. Between Edenhall and Warwick Bridge no less than 28,000 acres were enclosed. Such enclosure put the remaining open commons into even sharper relief. This part of England retains an important part of what may be considered a unique form of communal land management. The right of one or more persons to take part of the produce of the land belonging to someone else seems something of an anachronism in the 21st century.

Important flora and fauna conserved

In practice the retention of commons has conserved areas of open landscape and habitat of national and international significance. Within Eden there are many examples including the Crosby Ravensworth group of commons which is a candidate site as a special area of conservation under EU legislation. The flora and fauna of the heathlands, limestone pavements and water include rare and important features. Such areas need to be linked to the traditional practices under which they were created. Large areas of Eden are managed by farmers under agri-environment agreements on a scale matched in few areas of the country.

Even in the 19th century there were concerns about the disappearance of open commons under pressure by those wishing to enclose.

Writing his prize essay for the Royal Agricultural Society of England, Crayston Webster, the Westmorland land agent, noted: "If the remaining commons were where practicable enclosed ... better sheep might be kept and more profit made ... the only drawback is that we should more seldom enjoy a leg of four year old wether mutton: while the school of Lake poets and the shade of Wordsworth would doubtless pronounce it as a ruthless profanation, if their grand mountains were to be defaced by rigid lines of six foot walls set out by the surveyor's parallel ruler ... " [vi]

Powers of the manorial court

In earlier times the regulation of commons was the responsibility of the manorial court assisted by pinders, grassmen, reeves and hogringers. Using their devolved legal powers they were able through a jury of commoners to mete out local justice to those who transgressed. At its best the system was fair and equitable and capable of bringing all to account under the premise of good neighbourliness.

> We present Joseph Southwaite, the Vicar of Castle Sowerby, for surcharging (overstocking) the common with sheep against the custom.
> [Castle Sowerby Manorial Court, April, 1743, DMBS/4/16/CRO]

▲ *This unusual building, known as The Goose Egg, stood opposite Baron Wood farmhouse, Lazonby, and accommodated two families. It was demolished more than 20 years ago.*

▲ *Greenthwaite Hall, Greystoke, built by Dorothy Halton c 1650. The Latin inscription over the doorway reads: "Here we reckon ourselves pilgrims". The house stands only yards from Greystoke Park wall. Dorothy Halton, a thrifty farmer's wife, used to scatter oats on the lawn in front of the window to attract deer from the park. Sitting at the window above the doorway with her crossbow, she obtained a plentiful supply of venison.*

Cultural landscape — 27

▲ *The packhorse bridge, Ivegill (c1910), just wide enough for a packhorse. The low parapet prevents the pack being damaged.*

▶

Lanty trough in the wall of an outside loo at Ivebank, Ivegill. Farms which produced textiles needed ammonia for fulling the cloth. About two weeks prior to the fulling day, members of the family were instructed to fill the trough when visiting the loo. Receptacles were often distributed to obliging neighbours with a request not to eat cabbage since it affected the chemistry of the process.

In the management of open fell systems the system of sheep marking has remained an important and relevant means of identification. The sheep marks are collated in shepherds guides, a sort of yellow pages of sheep marks which date back to 1817. The four parishes from Milburn to Ousby produced a Cross Fell Goose Shepherds guide in 1862. The smit marks take the form of rings of coloured paint round the neck with permanent marks on the webs of the feet.

Other features of the landscape include the pollarded trees, particularly the ash which are maintained today as landscape features. From Viking times the trees provided feed for sheep "in winter to browse them with the tender shoots of the holly and ash." [vii]

Local styles of walls, buildings and sheepfold, together with the breeds of the area, shepherds meets and systems of husbandry, all combine to create local distinctiveness and diversity which need to be valued and sustained. This is as relevant to the Eden valley as in the surrounding uplands; the area needs to be cared for in a holistic and not fragmented fashion.

---

i Nicholson Norman The Lake District: An Anthology London 1977 pp 187, 188.

ii Webster C The Farming of Westmorland Prize Essay in the Journal of the Royal Agricultural Society of England 2nd Series vol 4 p 8.

iii Humphries A B Farming in the Lake District, an anachronism or the future of the Cultural Landscape in the Regional Bulletin, CNWRS, Lancaster University New Series No 13 p 20.

iv Directorate General for Economic and Financial Affairs Towards a Common Agricultural and Rural Policy for Europe Report No 5 1997, chapter 6.

v Webster op cit p 7.

vi Ibid p16.

vii Garnett The Agriculture of Westmorland 1800-1900 pp 16,197.

▲ The old farmhouse at Old Parks, Kirkoswald, c 1880. Old Parks was the special haven of the famous naturalist and broadcaster Romany (the Rev. George Bramwell Evans) who attracted huge radio audiences in the 1930s and 1940s. Many of his journeys in a gypsy caravan culmintated at Old Parks where his ashes were scattered and a memorial erected.

▲ Yeoman farmer clipping sheep in Hartsop. The spinning gallery is a reminder that adding value to primary produce was an integral part of the farm business until well into the 20th century.

▲ Toppin family graves at Skelton. The family were leading breeders of Shorthorn cattle and Oxford Down sheep.

Cultural landscape — 29

On Alston Moor, the restoration of the landscape at the former Thortergill Low Level lead and silver mine, at Garrigill, is a remarkable personal achievement over 30 years by Ian Johnston for which he was named *Countryman of the Year* by *Country Life* magazine. The eight-acre site is once again a wooded gorge through which Garrigill Burn runs open and unhindered. Above is Whitesyke Mine (c 1900) which operated in similar fashion to Thortergill, a short distance away. The picture on the left shows how Garrigill Burn was culverted and buried beneath mining debris when Mr. Johnston, a former Tyneside businessman, bought Thortergill which was owned in pre-mining days by his forebears. The debris has now all been cleared in the remarkable restoration which has produced a picturesque site (below) with tea garden and wrought iron smithy open to the public.

Cultural landscape — 30

▲ *Ancient cultivation terraces, known as lynchets, mainly found in East Cumbria.*

▲ *Distinctive architecture at Dale Head, Martindale.*

◀ *Haymaking at Ravenstonedale c 1910.*

▶ *Visitors and sightseers have long played an important role in the area. This picture, from about 1900, shows the Bishop of Durham on an Ullswater steamer. The little girl on the left was a visitor to Hause Farm, Martindale.*

▲ A country public house, The Fox Inn, at Ousby, about a century ago.

Unusual features in the landscape near Greystoke are farmhouse follies built by the 11th Duke of Norfolk in about 1789, with names from the American War of Independence, the British conduct of which he did not agree with. Above is Fort Putnam and left Spire House.

Cultural landscape — 32

One of the most distinctive inn signs in the district adorns the front of the Agricultural Hotel, at Penrith, depicting a scene typical of those which used to take place when the town's auction mart operated in adjoining buildings.

Shap Abbey, where white canons of the Premonstratensian Order produced and exported wool to Italian weavers in the 13th century.

Still just visible is the sign "Wilson's Guano Manure and Seeds Depot" across the gable ends of buildings in Cromwell Road, Penrith — a reminder of the importation from South America of guano — seabird droppings — which produced a dramatic rise in food production in the mid 19th century. Shiploads were transported by rail to Penrith from Whitehaven.

▲ *Martindale Church with Herdwick sheep and its ancient yew which pre-dates the 12th century building*

▲ *Mungrisdale sheepfold, the first to be built in a series by sculptor Andy Goldsworthy. This one took just four days and was completed on Friday, 12th January, 1996. One of the wallers, Steve Allen, lives at Tebay.*

Livestock

▲ *Thomas Bowness (left) and a colleague with horses and foals at Dacre Banks about a hundred years ago.*

FARMING IN EDEN is characterised by its livestock which are among the most interesting and attractive features in the cultural landscape, adapted to local conditions and the needs of the wider population. Like the landscape they change in response to the changing environmental and market conditions. Not only has the pattern of breeds altered over time but the breeds themselves also evolve and change.

From about the mid 1780s Britain became a regular importer of food and with a rapidly rising population, increasingly concentrated in urban centres, looking to the farming community to respond. Raising levels of production and productivity in Eden as in much of Britain signalled a need to improve livestock production through breeding and management.

In the early 19th century the stock kept on farms in Eden were closely adapted to the environment and farming systems. The emphasis in cattle was firmly on Longhorn and Galloway types. Local pigs were kept in small numbers and the sheep flocks were based on Herdwick and Blackfaced Heath types, with small numbers of Limestone Crag sheep and a native Greyface on the lower commons. The latter two breeds of sheep are extinct although the Crag sheep put on a creditable display at the Jubilee Show at Kendal in 1889.[1] The Cumberland pig also disappeared after the Second World War with changing market conditions which determine the profitability and ultimately the popularity of all breeds in the commercial farming sector.

▲ *Oxford Duke of Calthwaite 100th, the famous Shorthorn bull belonging to Mr. J. Harris, Brackenburgh Tower, leads a parade at the Royal Lancashire Show in 1928, the same year in which he was breed champion at the Royal Show, in Nottingham.*

O what a beautiful morning at Newton Rigg in the 1950s.

Through selective breeding and improved health and nutrition, dramatic improvements have been achieved particularly in the lowland sector. For hill farmers the need to accommodate the limitations imposed by the environment which includes areas of high rainfall and hard grazings continues to modify the ability to pursue the market entirely. During the life of Penrith Agricultural Society new breeds have been introduced with varying degrees of success. In the early 19th century the main impact came from the Improved Shorthorns and the Leicesters together with native terminal beef sires including the Hereford. Other have included the Ayrshire cattle and Southdown sheep popular in the mid 19th century for crossing on to draft fell ewes. Lincoln and Oxford tups and a whole range of continental sheep and beef cattle have been tried. [ii] Experimentation at farm level, allied to the application of science through sound advice, has made Eden an area recognised as a centre for quality produce within a landscape which is unique and of great beauty.

The pressures and demands of the market have undoubtedly diminished the number of commercial breeds which themselves in many cases have been changed by selection. Thankfully there are also those who have the commitment to maintain stocks of the minor and so called rare breeds. This should also be recognised as valuable. Diversity is a strength and does ensure wider choices in future.

During the 19th century through experimentation a stratified system of sheep breeding evolved which has been the hub of the industry over the last half century. Crossing draft hill ewes to produce "hybrid" ewe lambs as breeding stock for lowland farmers has underpinned the industry and ensured the lowland farmers supply of quality breeding ewes. The use of the fells as a reservoir of breeding sheep for upland and lowland breeders and finishers is a reminder of the links between farms from the fell tops to the valley bottoms. The crossbreeding of sheep in Cumbria has made it a national centre for Mule lambs. The sales at Lazonby, Penrith and Kirkby Stephen have been a focus for buyers from all parts of England.

The arrival of the Improved Shorthorn early in the 19th century allowed Penrith to become a national centre for the breed in the later 19th century until the 1950s. Just as the arrival of the Shorthorn led to the demise of the Longhorn in a few short decades so in the period after World War Two black and white cattle replaced the shorthorn domination which had lasted a hundred years.

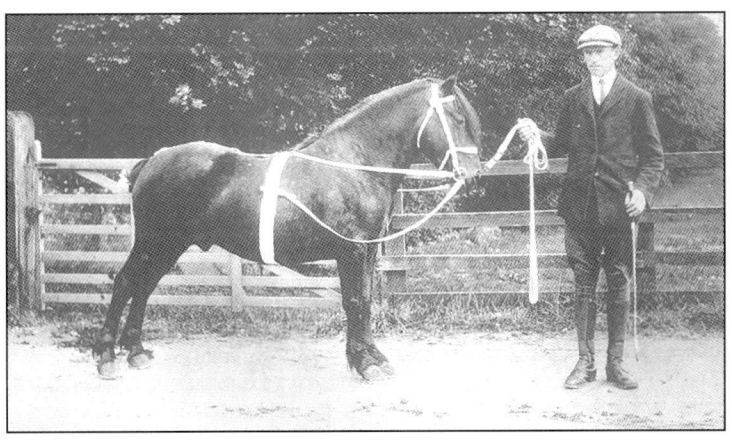

The famous Fell pony stallion Sir George, belonging to C. W. Wilson, Oxenholme, c 1910.

Clydesdale gelding "Clifton", President's Medal winner at Dundee Show, 1949. Owned by Irving Holliday, Clifton Hall, and bred by J. B. Milne, Lhanbryde, Elgin.

▲ R. W. Hawell, Threlkeld, with champion Herdwick ram Lord Paramount, in 1882.

▲ The celebrated Herdwick ram Nero, 1870.

▲ Cumbrian Shorthorns at the Glasgow Dairy Show in 1956.

▲ Forward creep grazing, a means of increasing sheep stocking rates introduced in the late 1950.

It is too early to evaluate the contribution of Eden breeders to the immense success of Friesian cattle; that will be for future historical reflection. Almost certainly a similar achievement will be identified.

In 1900 there were only about 30 or so Friesian herds in Britain, yet today they form well over 90% of the national dairy herd. During the 19th century large numbers were imported from Holland for slaughter but rarely for breeding. One of the celebrated foundation herds in Britain was that at Hawkrigg, near Wigton, established by John Twentyman in 1880 with five cows purchased as a result of a casual attendance at the sale of Mr C. W. Wilson, of Rigmaden, in Westmorland. The sire Royal Duke was perhaps the best known bull of the pre-registration period. [iii]

Three remarkable decades

The last three decades of the 20th century have been remarkable across the cattle sector. The introduction of Holstein blood into the dairy population has had a major impact on productivity. The early maturity and propensity to fat weighed heavily against some of our native terminal beef sires. The Charolais appeared in the sixties followed in the next decade by Simmental and Limousin and subsequently by others including the Belgian Blue and Blonde Aquitaine. The impact through improved growth rates, lean meat content and in offering new options in suckler cow breeding has changed the beef enterprises of Eden radically. The showground is also more interesting as a result.

Whilst the general pattern of the rise and fall of breeds is reasonably clear there have always been individuals willing to devote their allegiance to minority breeds. In the mid 19th century Mr Howard maintained a breeding herd of 200 gold and silver dun Argyllshire cattle in the park at Greystoke. "At a distance they resembled a herd of deer ... but they form a noble adjunct to the grounds such as no other park in the kingdom can boast." [iv] Through the 19th century a growing number of Irish store cattle came into and through the county, bringing Dexters, Donegals and Moyles. Around 1850 the Penrith Spring Fair for store cattle was said to be the largest in England.

▲ The champion Ayrshire at Penrith show in 1959, Winder Sweet Lavender, held by Anne Cottam, daughter of Mr. T. H. Cottam, Winder Hall, Tirril.

▲ County champion Wyandotte from egg laying trials of 1930. Full year's record: 329 eggs — special 139, first grade 124, second grade 62

The "Improved Shorthorns"

The primary interest in cattle has always been in the breeding. Perhaps the art is in anticipating future trends in relation to type. When the Improved Shorthorns began to establish around the time of the first Penrith Show there were real differences of opinion, with the Booth camp selecting primarily for beef and the followers of Thomas Bates pursuing milking and dual purpose objectives.

Those engaging in the shorthorn revolution took up the challenge with a clear leaning to the Bates blood. The first pedigree Shorthorn bred in Cumbria was Beauty at Lowther. The record appears in Vol 1 of the Coates herdbook 1822. [v] This proved decisive and by mid century the Penrith district was well supplied with stock of the Bates type. Son of the Gwynnes of J. C. Toppin at Musgrave Hall, Skelton, sustained yields of 30 quarts a day some considerable time after calving, being equivalent to a lactation yield today of 7,000 kg. Ewan Troutbeck's cows at Blencowe were exceptional for milk and flesh. Later at the Wreay, Thomas Richardson and from 1919 the Jacksons continued the standard. [vi] At higher altitudes W. J. Dent raised virile dual purpose types for their Southern buyers one of whom won the breed championship at the 1948 Dairy Show with an eight-year-old from Kaberfold. The bull, Kaberfold Viscount, won many honours including the Lowther Cup for pairs at Penrith in the late 40s before being sold into Wales for 400 guineas at the Penrith sale. Edward Jackson's herd, founded at Newbiggin House in 1931, included Thornby Foggathorpe 43rd which yielded over 2,000g in a lactation at 3.86% butterfat.

Farral, writing in 1874, offered the opinion that the best herd ever offered for sale was that of Mr Saunders of Nunwick Hall "which for blood symmetry and fashion are difficult to match". [vii] In the mid 19th century dairy cow yields averaged around 350g (1,600kg). By 1950, at the close of the Shorthorn era, it was 600g (2,700kg) and by 1990 had risen to 1,200g (5460kg).

In earlier centuries it seems that the Fell Galloway or Fell Pony was almost universal, being progressively replaced on lower ground with the larger breeds necessary for changing agricultural techniques. On the hills of Caldbeck, Bampton, Orton and Ravenstondale the local types bred " from time out of mind" continued to hold a significant territory.

Dominance of the Clydesdales

The major development came with the Clydesdale which was never seriously threatened by other heavy horses. Shires had a few adherents and the Earl of Lonsdale tried with the Suffolk Punch stallion Prince Albert which was offered for use in the district. Some

▲ Clipping day, 1919, at Coupland Beck, Appleby. On the extreme right is Jack Patterson, whose grandson, William Patterson, farms there today

use was made of the Cleveland crossed with the Clydesdale for carriage work but this declined after the coming of rail travel. [viii] The Penrith Show of 1844 included classes for Thoroughbred, Agricultural, Coaching and Clydesdale horses.

The Herdwick sheep is perhaps the quintessential Cumbrian native, with a history so rich and well known not to be addressed in this volume. Appropriately Penrith was the first of the local shows to offer prizes for Herdwicks, followed in 1844 by Appleby and Kirkby Stephen and in 1860 by Orton. In the 1850s the breed was to be found as far East as Orton, Tebay, Ravestondale and Kirkby Stephen, a territory later taken by the Roughs and more recently invaded by Swaledales.

▲ *Edward Capstick with Rough Fell ewes on the Howgills.*

By the end of the 19th century Herdwicks were already in territorial retreat. Garnett, around 1900, noted that standing at the head of Longsleddale the sheep to the West were all Herdwick whilst to the North and East were Swaledale, Rough Fell and Scottish Blackface. [ix]

In the debates that invariably accompany breed movements Johnathan Binns described Herdwicks as good milkers which seldom produce more than one lamb and are "terrible lish". All breeds have their descriptors which challenge an understanding of the English Language for non adherents. Of the many type statements for the Herdwick breed that of Joseph Hawell, of Lonscale, Threlkeld, in 1888 includes some interesting word pictures of the sort that have doubtless guided many show judges.

"The head legs of the lamb are mostly black but at three months old the nose and forelegs should begin to show grey hairs and increase in brightness until three years of age, when all brown, black or speckled marks should finally disappear, and the head, ears and legs become arrayed with pure bristles of a silvery hue, a coat consisting of a long strong broad staple, of thickly grown white wool, and supplemented with a blue or black flowing mane round the neck is considered of the highest importance. The head should be broad and a little Roman shaped and carried high, and adorned with a topping and bright prominent eyes ... the gait of the sheep should be wide and perpendicular with a prancing step and courageous bearing".

No wonder judging is so difficult!

Recognising the sensitivities, it is likely that Lowe, writing in 1842, in describing the "blackfaced heath breed" was assessing the progenitor of our horned hill sheep. He continued: "It extends across the vales of Kendal and Eden to the higher mountains of Cumberland and Westmorland on the West and by the Carter Fell into Scotland". [x]

By the beginning of the 20th century clear breed types had emerged as Swaledale, Rough Fell and Scottish Blackfaced. Among the early breeders of Scottish Blackfaced sheep was J. Irving, of Shap Abbey, who won the principal prize for the breed at the Manchester Royal in 1869 with further success at the Preston Royal in 1885. At Liverpool in 1877 William Hindson, of Sleddale Hall, Shap, won first prizes in both aged and shearling ram classes. [xi] The breed continues to hold a territory in the district on the East fellside and at Alston under the skilled care of the Walton family.

Breed recognition continued at least in part to be the subject of differing opinions. At an early meeting of Swaledale Breeders the members understandably rejected a proposal from Mr Dent, the joint secretary of the Blackfaced and Dales Bred Sheep Society, that the new organisation be called the Blackfaced and Dales Bred Swaledale Sheep Society. The proposal makes some of the EU directives look quite brief. Ultimately the title became The Swaledale and Dales-Bred Sheep Breeders Association. In 1923 the words Dales-Bred were deleted. [xii]

◀ *Rapt attention during the sale in 1962 of the noted Ayrshire herd of T. H. Cottam, Winder Hall, Tirril, which made over £13,000. Top price was 460gs for the senior stock sire, Pant Jewel Case.*

▲ *Mares and foals at Inglewood Inn, Stoneybeck, Penrith, in 1922. Farmer John Harrison and his son John on the left.*

The Rough Fell established the main part of their territory on the white ground of the Lune valley and have a long held reputation for quality wool for the mattress and carpet trade. It is significant that the Kendal town motto is "Wool is my bread".[xiii] In the wool records of Pegalotti, the Italian wool merchant in Rome, there is mention of wool imported from Shap Abbey which may well have been from the forerunner of the modern Rough. By 1848 the breed seems to have become distinctive. The Sedbergh Show differentiated between Blackfaced and Scotch in the sheep classes. This might have been the first "official recognition" of the breed which ultimately formed an Association in 1927. There are few more noble sights on a showfield than a Rough Tup.

Spectacular rise of the Swaledale

The Swaledale, claiming a distinct geographical territory from Kirkby Stephen eastwards down Swaledale as its genesis, has enjoyed one of the most spectacular rises in popularity in the last half century. The population of the breed in 1950 was 450k.[xiv] The breed reached 504k by 1971 and by 1996 had grown to an estimated 1.5 million.[xv] The Breeders association, formed in 1919, has been active and direct in its influence over the breed through registrations, shows and sales. Much of the C District falls within the Penrith area. Experience suggests that breeds need to move forward to maintain their position. Change can be seen as a threat. To ignore changes in the industry is even more risky. It is good to see that Swaledale breeders are actively involved in testing for scrapie resistance and are considering other facets of breed improvement.

A significant factor in the increasing popularity of the Swaledale was the demand for greyfaced or mule lambs sired by the Bluefaced Leicester. This proved to be an arranged marriage with excellent prospects. Jim Hall, writing in 1966, noted that in Cumberland and North Westmorland a developing feature of sheep production was the use of the Bluefaced Leicester to breed breeding gimmer lambs.[xvi] This proved to be another example of Eden farmers achieving national status as stockbreeders.

---

i Garnett F Westmorland Agriculture 1800-1900, Kendal circa 1910 p 170.

ii Dickinson W Farming of Cumberland Journal of the Royal Agricultural Society of England vol 13 1852 pp 207-300, and Farrall Thomas, Report on the Agriculture of Cumberland chiefly with regard to the Production of Meat, Journal of the Royal Agricultural Society of England Second series vol 10 1874 pp 402-429.

iii Hobson G British Friesian Cattle, Transactions of the Highland and Agricultural Society of Scotland Fifth series vol 30 1918, p 43.

iv Dickinson op cit p 253.

v Garnett F op cit p 185.

vi Burrows G T History of Shorthorn Cattle, London 1950 p 90.

vii Farrall Thomas Report on the Agriculture of Cumberland, Journal of the Royal Agricultural Society of England, 2nd series vol 10 p 409.

viii Garnett F op cit p 248

ix Ibid p 151.

x Ibid p143.

xi Ibid p 148.

xii Swaledale Sheep Breeders Association Journal (ed. Ruth Stephenson) Kirkby Stephen 2000, pp 61-64.

xiii Humphries AB Hill Farming in the Cumbrian Uplands A Cultural Perspective. Unpublished paper 1994.

xiv Hill Farming Research Organisation 2nd Annual Report HMSO 1953.

xv Anderson J 1.4 million and rising, Swaledale Sheep Breeders Association Millennium Journal. Kirkby Stephen 2000, p 23.

xvi Hall J S Hill Farming in the four northern counties of England. Journal of the Royal Agricultural Society of England vol 127 1966 pp 17-28.

▲ *Mr H. G. Marshall, Croglin, won the Swaledale championship at Kirkoswald show in 1959 with a two-shear ram.*

▲ *Inglewood Molly Millicent, winner of five championships and 33 other prizes at the Royal and other shows. "The heroine of the Inglewood herd ... no more magnificent or complete type of true milker was ever seen ... " — Frederick Punchard, addressing the Royal Association of British Dairy Farmers, 1892.*

▲ *Edward Wood, Cooper House, with the first prize Blackfaced Rough Fell ewe, Kendal, 1908.*

◀ *Edith Dent holds the young bull Kaberfold Prince 8th which won the Shorthorn championship for her brother, John B. Dent, of Kaber Fold, Kirkby Stephen, at Kirkoswald in 1959.*

▶ *Winning Shorthorns from Crossgrigg, Cliburn, at Penrith Show on the Foundry Field in the 1950s.*

▲ Members of the national Sheep Breeders' Association on a visit in 1955 to the farm of Richard M. Wilson, Glencoyne, Ullswater, a noted breeder of Herdwicks who is holding a picture of one of his prize rams. The ram held by E. W. Tyson, Grasmere, is "Dick's Permission", so called because it was the result of a negotiation with Mr. Wilson for use of his ram.

▶ Newton Rigg Conductor, which won first prize at the Royal Show on 1953. The picture was taken in the Station Field, Penrith.

◀ Limestone Crag sheep, now extinct, and formerly found on the limestone areas from Bampton to Arnside.

The Cumberland pig

An impressive example of the Cumberland pig.

FAITH, Hope and Charity were the first prize sow pigs under six months at the Carlisle Royal Show in 1855. Faith was out of a sow bred by a Mr. Unthank, of Netherscales, Hutton-in-the Forest.

The names drew comments, as might be expected.

"And, pray, which of these is Charity?" said an old lady after duly adjusting her spectacles and taking a protracted survey of the trio.

"Which is Charity marm?" said the attendant. "Why the biggest on 'em is Charity.

"My dears," said the old lady, turning to her daughters. "I never saw it put in that practical way before."

Source — H. H. Dixon, The Druid.

Dainty Dinah, winner of the British Dairy Farmers Association prize for the best milk recorded dairy cow at the Penrith 1924 show, for J. Smith, Far Close, Knock.

An early 20th century Blackfaced Swaledale ram.

▲ *Joe Norman, Skiddaw House, singing to the assembled company at Skiddaw Shepherds' Meet, at Wyllie Ghyll on the last Monday in July c 1910.*

▲ *Fell ponies at Bampton c 1910.*

▲ *Heavy horses breeder and exporter Irving Holliday, Clifton Hall Farm, near Penrith, pictured in 1967 with two 18.1hh Shire geldings destined for Alberta, Canada.*

▲ *Herdwick champions at Keswick Tup Fair, 1967.*

Livestock — 45

▲ The sheepdog can rest while its masters get on the with the hand shearing.

▲ A young visitor in 1950 to Low Baron Wood Farm, Armathwaite, has his hands full with quad lambs born to Tot and Bob Faulder's ewe.

Rural life and the farming year

▲ The beginnings of a mechanical revolution on Cumberland and Westmorland farms. The picture is believed to be of the first tractor ever seen in the two counties being demonstrated in 1905 at Dalston, on the farm of Mr. Robert Tinneswood, who is seated on the self-binder.

THE ONLY CONSTANT IN EDEN'S landscape is change, reflecting the seasons, the farming calendar and rural communities. Seen over the life of the Penrith Agricultural Society they have been dramatic but not all pervading. Many elements which in their time were perceived as permanent and vital have gone for ever. The working horses, droves of cattle and the army of workers are but a memory. Many traditional crops are absent from much of the landscape. Seasonality has been modified with the grass harvest running from May to October. For those remaining in farming there is little opportunity to draw breath between the seasons.

Yet the beauty of Eden remains and new features appear in place of the old. The quad bike with resident dog, travelling gangs of contractors bringing the latest technology to support a workforce reduced to a minimum and new breeds of livestock all add to the cultural landscape. The fundamental truth remains that farming families are not in the landscape, rather they are part of it.

"The old countryman is made wise by the wisdom of the earth. His daily contact with the principles of growth, his continuous traffic with the seasons may leave him subject to superstitions; but superstition itself is an acknowledgement that there is a mystic something behind phenomena"
— Sir William Beach Thomas, A Countryman's Creed, Cumbria.

Perhaps the most disconcerting and discernable element of change is the decline in the farming population. The relentless replacement of people with capital and machines has ultimately changed rural society fundamentally. Yet it could be argued that even more remarkable is the colour, tradition and cohesion of rural life that has survived. Seen objectively, the changes in population distribution between urban and rural England might have been expected to marginalise the cultural and society as well as the economic contribution of the countryside.[1] Despite the marginalisation of food production as an economic activity, rural life is more resilient.

▲ School gardens were important practical aids to learning in rural communities. These boys were pictured in Askham School garden in 1938.

Rural life — 47

▲ *Domestic science teachers at Newton Rigg in 1938 on the first course in Britain to promote the use of British farm produce.*

▲ *The art of "drenching".*

▲ Families listen to some finer point about horticulture on a visit to Newton Rigg Farm School, c 1950.

"... so much of the past of the country, its feelings and its literature was involved with rural experience ... that there is almost an inverse proportion in the twentieth century, between the relative importance of the working rural economy and the cultural importance of rural ideas." ⁱⁱ

Men, horses and machines

At the time of the first show the general source of power was provided by horses fuelled by vast acres of oats and managed by an army of people. Although steam power became a factor in the mid 19th century its impact in Eden, dominated by pastoral farming within a framework of small fields, was not great. In 1867 Messrs Nicholson, of Kirkby Thore, were using a steam 8 horse power, 3 furrow plough which could also be used with a cultivator and drag harrow and used for threshing. Steam threshing machines began to come into use around 1860. ⁱⁱⁱ In the Penrith area the names of Messrs Dowthwaite and Stalker are particularly associated with the system.

More relevant to typical farms was the development of horse drawn machinery though there was a period of experimentation and testing which identified the need for significant modification. Many of the national shows featured demonstrations, tests and evaluations for the benefit of potential purchasers. In 1861 Admiral Elliot, agent to Appleby Castle Estates, bought a two-horse mowing machine at the Royal Show Leeds for use on the home farm, one acre per hour being very well cut with it. McCormicks American reaping machine was tried at Boustead's farm, Hackthorpe Hall, near Lowther, on September 20th, 1851. The day being dry it acted well but the driver would not cut downhill with it. Uphill it cut very well, drawn by two strong horses. These were changed twice in four hours during which period a little over two statute acres were cut. ⁱᵛ

▲ Hedging on the Newton Reigny road in 1934.

▲   *A Shorthorn study in the stackyards at Newton Rigg.*

A horse to every worker

By the 1920s the ratio of horses to people was just under one, giving a power per person engaged in farming of literally one horse. By the 1980s the ratio of tractors to people in farming exceeded one to one, giving a power available perhaps 80 to100 times greater. [v]

Gone are the numerous ploughing teams turning over a few acres a week, the groups of turnip singlers, families stooking sheaves following the binder. Now the forays of machines into the field are brief and with few people to be seen, stock inspection is now done with the aid of a Land Rover, pickup or ATV. The teams of hand milkers have given way to the technology of the parlour and lone herdsman.

Gone too is the 1947 Agriculture Act born out of wartime vulnerability to blockades twice in three decades. Through guaranteed prices and assured markets the legislation had clear aims to promote and monitor a "stable and efficient industry, capable of producing such part of the nation's food and other agricultural produce as in the national interest it is desirable to produce in the United Kingdom". [vi] Since then the CAP or Common Agricultural policy has taken over and has been shown to be incapable of adjusting production or even limiting expansion. Price constraints, quotas, and decoupling of support from commodities have been progressively implemented over the past two decades but with little encouragement to farmers. Agriculture is once again at the crossroads with deep implications not only for the farming community but also to the ancillary services and ultimately the landscape and environment.

▲   *A well earned tea in the harvest field around the turn of the century.*

Rural life — 50

▲ *The Mill Inn, Mungrisdale, in about 1900.*

Observed changes

1861

" ... I hired to serve Mr Abbot of Thornthwaite Hall ... there were 200 acres of hay to mow, and it was just one fine week ... there were two men who were paid so much to mow by the acre. I was sent to Clifton with a horse and cart to bring them. They brought two barrels of ale or old hock with them together with their scythes and tackle. They ld a glass and we had to call at every public house we came to and we all had more to drink than was good for us ... my head being not very clear ... the cart went right over and the two barrels of ale went rolling down the hill until a fence stopped them ... the men I brought cut 96 acres ..."

"the spring being so bad and thousands of sheep were lost owing to the snow and bad weather, it was a heartbreaking time for the farmer."

"In the autumn they that could soap (salve) their sheep commenced. Their wool had to be shaded a little distance apart with the fingers and thumb, about three inches apart, and the soap put on to the skin with the two forefingers. The soap was composed of tar and firkins of strong butter, unfit for family consumption." vii

1920/1

"Harvest time was here. We had no self-binder. We used the mowing machine to which, behind the cutting knife, was attached a rack of sorts. John Shuttleworth sat on the seat with a rake in hand. I led the horses ... when the equivalent of a sheaf had fallen on the rack John ... pushed it off to the ground. His daughters and the lad followed round... binding the sheaves with stalks drawn from them. It was slow and laborious work ... in due course when sun and wind had done their work the sheaves were carted ... and safely stacked ... inspiring in us the appropriate mood for the Harvest Thanksgiving service ... soon the pheasants and crows and a host of other birds found rich pickings in the stubble."

"During the winter the threshing machine came, a veritable monster ... towering high, emitting volumes of steam and smoke, shaking, rattling ... with a laboured cacophony. The men stood on the monster seeming (to be) its servants urgently and industriously feeding it ... The bolder spirits among the hens had a field day filling their crops with the scattered grain."

1922/3

"Late June found us in the hay. Mowing machines apart, machinery was non existent and we became masters of fork and rake, turning, shaking , rowing ... And the moment when tired but content we brought the last load into the barn, what a sense of fulfilment was ours ... when I had mown round the edge of the fields with my scythe I knew haytime was really over." viii

"I spent some time at the grindstone sharpening bill hook, shears and slasher , for it was now time for hedge trimming or slashing ... A year's growth had to be cut away ... muscle and sweat, hour after hour, this was the order of the day, as was the raking up the cuttings for burning ... I felt rewarded as I stood back and surveyed my workmanship, the hedge neat and tidy. ix

▲ *Threshing day at Row End Farm, Warcop, c 1902. Thornborrow Richardson, the tenant farmer, who also had a butchery business, is standing on the right with a rake. The burly figure with the black hat second from left is his brother, Ned, who was an auctioneer at Lazonby for Penrith Farmers. The threshing outfit belonged to Hullocks, Warcop.*

"When roads needed to be made up, cartloads of stone would be brought from the nearest quarry. These would be put in a large heap, often six or seven yards in length on the verge by the roadside. The stone breaker held a very large iron-headed hammer in his hand and with this broke the stones. Holding each piece in position with his left foot he would gradually work through the pile. Sometimes another man would work with a smaller hammer reducing them further for laying raking and finishing." [x] In areas of high rainfall the maintenance was a labour intensive business.

Extracts from the Surveyor of Highways Account Book, Mungrisdale:

```
1838
Dec 24th Lot 16 Breaking 9 yds of stone at 1s 3d
Rich. Greenhow           11s 3d
1840 Blake Beck Bridge built in September Richard Hebson
public letting           £6 15 0d
1841
January 6th Seven and a half days snow cutting at 2s 0d
Jos Bowerbank            15s 0d
```

"Some dalesmen's houses have the dwelling-house barn, and ... cowhouse all under one roof. Sometimes the central part was a sort of passage with doors leading on one side to the habitation of the family, and the other to that of the animals together with a barn. The reason we call the entrance to a house the threshold is because the threshing floor was placed there. In the dalesman's house he used to feed not only his family, but his labourers who arranged themselves according to seniority at a long table remote from the fire, while he and his family sat at a round table near the hearth." [xi]

"Washing lasted all day; the water had to be carried from the pump and heated in the boiler, then the clothes were washed, then boiled, then steamed, next they were put in a bath of cold water with Reckitts Blue. The only soap was Hudson's Powder, and soda was also used. It was really hard work. The old wooden roller mangle was kept outside the door as it was too big to come inside. I can still recall having to stand out there on a bitter cold day and get all that washing through that old wringer." [xii]

Responding to change

"A better education for the rising population. In dealing with the various manures now pressed on his notice ... the farmer has in some measure to grope his way in the dark. It is impossible that all farmers should be chemists and geologists and botanists but it is certain that the rising generation would benefit by more knowledge of these subjects than that possessed by their forefathers." — (Crayston Webster Westmorland land agent 1868) [xiii]

Although Penrith Agricultural Society was in place for 60 years prior to the establishment of a farm school or institute at Newton Rigg, the importance of education had not only been voiced but demonstrated in the years before 1896. The experimental and demonstrational work by local landowners and the formation of agricultural societies and debating groups were quintessentially educational in concept. As agricultural science matured, the need to underpin practice with understanding grew. Local leaders including Harry Howard, of Greystoke, in Cumberland, and Frederick Punchard, the Underley agent in Westmorland, looked for an opportunity. Fortuitously, this came in an unexpected and unplanned manner. In the late 19th century the temperance movement was lobbying government to reduce the number of licensed premises in the countryside. A bill was introduced to parliament to impose additional duties on whisky and beer with a view to compensating those who lost their licences. When the intention became known there was a public outcry against using the income in this way. In a lethergic and half empty House of Commons A. H. D. Acland proposed the use of the money for technical education and rate relief.

Around 12 per cent. of the fund was allocated to agricultural education by local authorities. The twin counties were among the most responsive in the country, perhaps benefiting from the experience of the pioneer independent college at Aspatria founded in 1874. Some Penrith students had experienced the provision at Aspatria which established a national reputation for combining science and practice.

▲ *Mechanical harvesting, but still relying on horse power.*

Farm school pioneers

Both H. C. Howard and Frederick Punchard had been associated with the Aspatria college which both counties encouraged. When the Newton Rigg college was established as the Farm School, the Cumberland and Westmorland Advertiser led with a headline "An Agricultural Institution for the Sister Counties." xiv

William T. Lawrence, who had been delivering classes and lectures in Cumberland for the County Council, was appointed Resident Manager at an inclusive salary of £300 to include the services of his wife as matron. A weekly allowance of 8s per head was made to him for feeding each of the students in residence. xv

The appointment was not only critical but highly successful. W. T. Lawrence was a natural communicator with students and farmers, demonstrating an empathy with rural communities and great skill in promoting modern agriculture.

By 1900 Mr Lawrence reported that Miss Armstrong, the dairy instructress, had judged butter at nine shows. At the 1899 Penrith Show Newton Rigg had a display of dairy produce. Fresh, week old and packaged butter was on show together with three kinds of cream cheese, fresh cream, cream in stone jars and Cheddar cheese. All of the April and May cheeses were taken from the show by Grahams who established a substantial business in Penrith as factors of dairy produce.

A key foundation of rural life

Following the First World War the contribution of women to rural life seemed to gain recognition. Lord Henry Bentinck, of Underley, was a prime mover to establish the Women's Institute movement in the two counties. The Westmorland Federation came in 1919 with Cumberland following the next year. Lindsey Robb, the Newton Rigg Principal, convinced of the educational potential, urged the Joint Committee to make every effort to work with the WI. The movement proved to be one of the key foundations of rural life and continues to contribute locally and nationally. The Young Farmers movement was established nationally in 1921. Early clubs included Great Asby, Bolton and Alston. J. H. Faulder addressed a meeting at Newton Rigg in 1937 when the Cumberland and Westmorland Federation was formed and the Newton Rigg vice-principal, Arthur Mann, made secretary. Newton Rigg helped to foster the movement in association with the many supporters from the farming community. The concept of Young Farmers proved more difficult to communicate to the wider community and frustrated the efforts to gain County Council funding for staff. C. H. Roberts at a County Council meeting observed that many councillors thought that these clubs were where young farmers went to play billiards. Mr Marrs was more direct suggesting that they were for the children of "toffs" and that they could be described as "pairing off clubs".

▲ *Ploughing on Kirkstone Pass in 1940 — a response to the Dig for Victory campaign.*

▲ *Horse and horsepower in harmony in the late 1950s, sowing and harrowing in grass seeds in a field at Carleton, near Penrith, now part of a housing estate. Behind the Clydesdale pulling the drill is William Young, who farmed at Longacres, Carleton, for 40 years. He was doing the sowing job for Joe Kerr.*

Further opportunities to develop the well-being of rural communities lay in the schools. In 1911 Newton Rigg was working with 31 schools with gardens. By 1932 the figure was 64 and in 1940 over 100 school gardens were in use to demonstrate principles which could never be achieved in the classroom. Some experimental work was undertaken in school gardens where the competence of the teacher was high. Co-ordinated trials with soft fruit, potatoes and carrots contributed to the extension programme in the two counties. From 1929 schools were used to deliver evening classes at four centres in Westmorland. In 1904, when the rural economy was depressed, Greystoke School requested the loan of fruit trees and bushes for the garden. The idea of returning them 20 years on is interesting!

Further links with schools came with courses for teachers dating from the early 20th century. In the depression of the thirties Newton Rigg ran the first course nationally for domestic science teachers specifically to promote British Farm Produce. [xvi]

Taking a national lead

Board of Agriculture and Fisheries Leaflet 275 (1916):

"A satisfactory way of eradicating bracken is to switch off the shoots ... with a piece of fencing wire twisted on at the end of a stick. A number of boys under proper supervision could be used for the purpose ..."

Board of Agriculture and Fisheries Leaflet 277 (1914):

Tuberculosis in Farm Stock. "The disease is contagious and affects man, many species of mammals and birds ... probably not less than 25 per cent of adult indoor cattle ... are affected."

Board of Agriculture and Fisheries Leaflet 300 (1915):

Breeding of Useful Pigeons. Until recent years the supply of table pigeons ... was practically dependent on imports ... Italy being by far the largest producer ... English breeders ... have given considerable attention in the last decade to the improvement of table pigeons ... as a result there are now many British lofts devoted exclusively to the rearing of table pigeons on commercial lines.

...well grown squabs are ready for killing when between four and five weeks old ... the weight of a Mondian squab should be from one to one-and-a-half pounds ... in some districts dealers prefer to buy live pigeons ... Only birds known as "squeakers" i.e. those which still utter the nestling note are suitable for this market, a point the intending breeder will do well to bear in mind.

Rural ministry

The image of sacred places as an integral part of rural England has been more frequently articulated than might be expected in a highly secularised society, even in rural Eden.

Fresh fields and woods, The Earth's fair face,
God's footstool and man's dwelling place,
I ask not why the first Believer
Did love to be a country liver. [xvii]

There is something of Englishness and spirituality which is perceived by many as more tangible in the country. Some forty years ago the biographer S. C. Carpenter wrote that "in the countryside somehow God seems nearer". [xviii] Along with the school and pub the church today still represents an anchor point in the rural landscape seemingly unchanging in a world of change. Even for non-Christians this can hold true; the rural writer Fraser Harrison, a professed non-believer, offering the insight that "the church moves me too ... because it unifies the village and its landscape in a single reciprocal creation". [xix] In many Eden churches and chapels the high point in attendance is Harvest Festival which itself provides a link between Christian liturgy and the festivals and rites of the pre-Christian era. The dressing of churches and chapels using harvest loaves, sheaves and corn dollies reflects a preoccupation with our farming past.

"The folds shall be full of sheep: the valleys also shall stand so thick with corn that they shall laugh and sing". (Psalm 65)

Monastic lands of the North

Farming and the churches have a long association. The tithe issue was but one element. More positive was the change wrought following the Norman Conquest when many of the orders seeking solitude found ideal settings in Northern England. Of equal importance they saw the potential to clothe the fells with sheep to produce crops of wool which would finance the building of the wonderful monastic churches which today many admire but few associate as a product of agricultural prosperity in the 11th to 13th centuries. The white canons of the Premonstratensian Order found the solitude they sought at Shap by 1120 [xx] and engaged in the production and export of wool to the Italian weavers. At that time the wool of 6-8 sheep would pay a shepherd's wage for a year. Other monastic lands included Gaythorne Hall in Asby parish, which is thought to have belonged in the pre-reformation period to the Cistercian House of Byland, one of the three luminaries of the north. [xxi] Other monastic properties included land given with Crosby Ravensworth Church to Whitby Abbey in 1109 and St Mary's, York, with lands at Meaburn and Wetheral. [xxii]

▲ *After the Second World War fertilisers played a key role in raising production.*

A tradition which has survived in some of the North Westmorland communities is that of rush bearing. In Eden, Great Musgrave and Warcop actively keep the tradition alive. Until perhaps the 18th century the floors of churches were of rough flags or beaten earth. Covering them annually with rushes was the equivalent of laying a carpet. At St. Columba's Warcop the festival is linked to St Peter's Day when girls carry crowns or garlands and boys carry rush crosses. The procession, headed by a band, visits the Lord of the Manor and then to church for a special service. The rush garlands and crosses are presented at the altar and then hung on the walls for the year. A special tea and sports come next, after which the traditional dance has in recent years been upstaged by a domino drive. [xxiii]

Memorial to "veterans of the chase"

An unusual memorial can be found in the Northeast corner of St Mary's Churchyard at Threlkeld where "to lift up mine eyes unto the hills" seems the most natural thing to do. The monument which stands about seven feet high includes the following information inscribed into the slate.

> "Around them stand the old familiar mountains, awake to drowning echo and confound, their perfect language in a mingled voice.
> The f(a)rest music is to hear the hounds rend the air with a lusty cry."
>
> **A few friends have united to raise this stone in loving memory of the undernamed who in their generation were noted veterans of the chase, and most of whom lie here buried in this churchyard.**
>
> **John Porter, of High Row, died May 4th, 1894, aged 48 years — 26 years the Huntsman of the Blencathra Foxhounds.**
>
> **1843-1903 John Crozier, The Riddings, died in March, 1903, aged 80 years — 64 years the beloved Master of the Blencathra Foxhounds and mainly through whose instrumentality this monument was erected.**

▲ *Stanley Harrison, Midtown Farm, Clifton, ready to hand milk in the kit required by regulation in the mid 1930s, with brother John.*

▲ *The winter of early 1963 was one of the worst of the century and some farms, cut off by snow, ran low on foodstuffs. This view is from an RAF helicopter as it descended with emergency supplies for Wrenside, Kaber, where Fred Ellwood was the farmer.*

Farming folk are perceived as having the capacity for self reliance and thrift. John Robinson, the Rector of Clifton, provides a somewhat similar picture in the case of a Patterdale curate, the Rev. Mr. Mattinson.

"*He buried his mother; he married and buried his father; he Christened his wife, and published his own banns of marriage in the church; and he Christened and married all his four children, a son and three daughters. He died January 31st, 1766, at the age of 96 years, 60 of which he had been curate of Patterdale. Till the last years of his life his stipend did not exceed £12 and never reached £20 per annum; yet such was his industy and domestic economy that on this small pittance he contrived to live comfortably and to save a thousand pounds.*" [xxiv]

Perhaps conditioned by the tithe demands, some farmers engaged in gentle repost with the clergy. William Dickinson records this encounter between shepherd and parson in Martindale:

> **The Two Shepherds**
>
> During sultry weather, the mountain sheep are very subject to be blown by the maggot fly, and they require to be closely looked after, to prevent losses in the flocks. One Sunday a Martindale shepherd was returning from his duties on the fell when he encountered the parson going home from church service.
>
> Parson: "Well John, I suppose you have been attending to your flock, as I have been to mine."
>
> John: "Ey, as hev. And hed you any in't whicks*?" [xxv]
>
> (* whicks are maggots)

▲ *Winter 1940 and John Harrison, Midtown Farm, Clifton, sets off to fother sheep on the moor.*

i Short B. The English Rural Community, Cambridge, 1992, p 1.
ii Williams R. The Country and the City, (1975 edn.) p297 quoted in Short B op cit.
iii Garnett op cit p 206
iv Ibid p 205.
v Blaxter K. and Robertson N. From Dearth to Plenty, The modern revolution in food production, Cambridge 1995 p57.
vi Ibid pp 22, 23.
vii The three extracts are from the journal of Thomas Irving Collected by his son Timothy J. Irving. A copy is deposited in the Record Office at Carlisle.
viii Rememberable Things, Journal of a Westmorland Farm Worker and Methodist preacher Stanley Finch, author's collection.
ix Ibid.
x Street S Ed. A Remembered Land, Recollections of Life in the Countryside 1880-1914, London, 1994 p 112.
xi Ibid p151.
xii Ibid p156.
xiii Webster C. The Farming of Westmorland, Journal of the Royal Agricultural Society of England, second series vol 4, 1868 pp 36, 37.

xiv Cumberland and Westmorland Advertiser, March 3rd 1896.
xv Humphries A.B. Seeds of Change, Penrith 1996, p 30.
xvi Ibid p 55.
xvii Vaughan H. Retirement, in L. C. Martin Ed. Henry Vaughan: Poetry and Selected Prose (Oxford 1943) p 436.
xviii S. C. Carpenter Winnington-Ingram (1949) p 349.
xix Ibid p165.
xx Butler L and Given-Wilson C, Medieval Monasteries of Great Britain, London 1979 p 344
xxi Ibid p165.
xxii Personal communication from Mr J. T. Relph, author of The Chronicles of Crosby Ravensworth.
xxiii Notes provided by Michael Gregson of Warcop.
xxiv Taken from a Guide to the Lakes 1819 and quoted in Nicholson N, The Lake District, An Anthology, London 1977. pp194, 195.
xxv Dickinson W. Cumbriana London 1876 p176.

Rural life — 56

▲ *Visitors from Glasgow enjoy helping the Fisher family with the haymaking at Moss Thorns Farm, near Penrith, just after the war. A cut-down lorry provides the motive power.*

▲ *The advice in 1939 for dealing with a possible gas attack involved oilskins and respirator for the man and sacking to protect the eyes, nose and mouth of the horse.*

▲ *Rogation Sunday at Bampton in 1965. Crops, animals and machinery were blessed by the Vicar, the Rev. G. A. de Burgh Thomas.*

▲ *Firemen tackle a hay barn fire at High Field, Tirril, in the 1960s.*

▲ A royal occasion at Ousby.

▲ Another role for a farm horse at High Hesket — the Coronation celebrations of 1953.

Rural life — 59

▲ *Mother and son off to market in Penrith from Inglewood Inn farm in 1929.*

▲ *Hens were once a feature of every farmyard. This picture of John Harrison and a maid was taken at Inglewood Inn in 1922.*

▲ Orton Young Farmrs Club members showed their acting skills in 1959 when they entered two plays in a county festival — a comedy, "The Bathroom Door", and a thriller, "Immortelle".

▲ A group of trophy winners at Ullswater Sheepdog Trials in 1959, with the President, the Earl of Lonsdale, and the Countess. The committee chairman, Mr. J. V. Allen, is on the extreme right.

▲ *Rushbearing at Warcop before the First World War and Cumberland and Westmorland style wrestling at the sports which formed part of the celebrations.*

Rural life — 62

▲ Young farmers pictured at their annual ball in Penrith's Crown Hotel c 1949.

▲ Margaret Wales, Thackwood, Raughton Head, didn't forget her riding pony Twig when she married in September, 1965. Twig enjoyed a large slice of wedding cake from Margaret after the reception.

Rural life — 63

▲ Penrith area YFC members enjoying an evening at the Cosmo night spot in Carlisle soon after it opened in the 1960s.

▲ Penrith area young farmers entertain in a concert in St. Andrew's Parish Rooms, Penrith. In the foreground are Jean Gowling, Marjorie Errington and Amelia Bowness.

▲ Colin Fawcett, Pembroke House, Brougham, who later became a barrister, performs in the YFC concert in St Andrew's Parish Rooms.

▲ A century ago rural communities did not have far to travel to find well stocked shops, as this photograph of Hilton and Co., grocers and drapers, in the village of Brough, testifies.

▲ Tractor ploughing in the post war period revolutionised work rates on farms.

▲ Brough garage owner Watson Sayer hired a Shorthorn bull from a local farmer in November, 1959, to greet Westmorland County Council workmen and planning officers when they arrived to demolish an illuminated sign which had been erected without planning consent.

▲ In splendid decoration, Jonathan Harrison's team from Anchor Farm, Penrith, competes in a ploughing match at Newton Rigg.

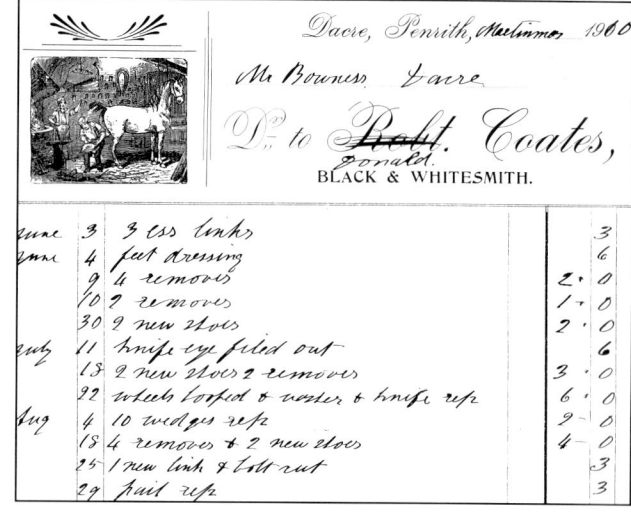

▲ A 1910 horse shoeing bill from Donald Coates, black and whitesmith, of Dacre.

◀ The WARAG letter to farmers notifying them of plans to use land in North Westmorland, centred on Lowther Castle, for tank training.

▲ Country life comes to town. A meet of the Ullswater Foxhounds, starting from Penrith's Royal Hotel, in Wilson Row, was an annual fixture when this picture was taken in about 1960. Huntsman Joe Wear, his whipper-in and supporters enjoy a stirrup cup provided by landlord Jack Walker.

People

PROGRESS IS NOT AN ACCIDENT BUT A NECESSITY [i] – these are the words of Herbert Spencer the philosopher whose life was broadly contemporary with the first 70 years of the Penrith Agricultural Society.

A major objective of agricultural and show societies has been to demonstrate and bring to a wider audience of farmers examples of good practice, changing techniques and livestock improvement. Clearly, progress depends on many factors which include research, education, advice and demonstration followed by more general adoption. Progress depends substantially on the leadership of those with vision, energy and commitment. Progress in a wider sense needs to confront and resolve those issues which constrain and damage farming. Underlying the whole spectrum of change is the need for practitioners, whether farmers or workers, to translate opportunity into practice. Of no less significance is the work of those who engage with rural communities in a social sense ensuring that they contribute to the well being of Cumbria and have access to an appropriate range of services. It is clear that rural development has always depended on each part contributing to the whole through the actions of people.

Classical accounts of the radical and revolutionary changes of the late 18th and 19th centuries highlight the contribution of a few heroic men like Bakewell, Townsend, Coke and Sinclair. To these without apology can be added the name of William Blamire of Thackwood (1790 = 1862), Tithe Copyhold and Inclosure Commissioner, not only as a contributor to the techniques of improvement but more especially for selfless public service. A nephew of John Christian Curwen and a regular visitor to the experimental farm at Schoose William undertook his own share of experiments on his estate inherited from Mr J. Sanderson, a relative from Plumpton.[ii] At the first Penrith Show William Blamire spoke at the dinner.

Solving the tithes issue

His real contribution however lay off the farm, advocating and acting to resolve key national issues. Of these the Tithe question was of particular concern and had defeated the best efforts of both parties in parliament. Wm Blamire successfully won a place in parliament and in his victory speech outlined his personal manifesto. " I hope that a reform in parliament is a prelude to a reform in the church. I believe tithes to be the most improper, the most iniquitous mode of payment ever devised by the ingenuity of men". [iii] Tithes, a tenth part of the produce of the land or the stock on the land, were resented by farmers and the cause of deep seated differences with tithe owners. This may be a reminder that even in the perceived arcadian world of the early 19th century farmers were wrestling with legislation and regulation.

Despite several unsuccessful efforts by both parties, progress seemed impossible. On Friday March 25th, 1836, Wm Blamire made the definitive speech on the subject which led to the Tithe Commutation Act 1836. The speech had impressed the House and Blamire received congratulations from the leaders of both parties, Russell and Peel. [iv] Recognising Blamire's ability and his understanding of the issues born of experience, he was appointed the first Tithe Commissioner for England, working from Somerset House. The task to convert tithes and tithe payments to a fair and equitable rent charge was immense. Some 14,829 districts were identified for each of which a tithe agreement or award was required. [v] Wm. Blamire's daily post on the subject amounted to 300 letters. Once agreements were in place maps and documents were required. The production of maps prior to the work of Ordnance Survey was in itself a Herculean task. The last surveys were complete by 1866. From 1845 Blamire held the three appointments of Tithe, Copyhold and Inclosure Commissioner simultaneously.

The Blamire Prize

William Blamire is remembered as a kind obliging man always ready to serve in the interests of agriculture. On his death a number of friends locally and in parliament subscribed to a monument in Carlisle Cathedral and with the remaining funds established the Blamire Prize which takes the form of a medal. The award has been used in a variety of ways but is now used to recognise service to agriculture in Cumberland. In the Millennium year it is appropriate that a recipient was Ruth Morton for service to Young Farmers Clubs.

▲ *Henry Charles Howard, Greystoke, a prime mover in agricultural education.*

▲ *Sir Jacob Wilson, Director of the Royal Show from 1875 to 1892, who was born in North Westmorland,*

Richard and Ann Harrison. In 1850, the year after their marriage, he purchased Thornborrow's Auction at Canny Croft, Penrith, one of the earliest auction marts in England.

Sir Jacob Wilson, 1836-1905

Jacob Wilson, one of the best known and popular agriculturalists of the 19th century, was born at Crackenthorpe Hall, the son of a local farmer. After a distinguished education at the Royal Agricultural College where he held the post of honorary farm bailiff whist in attendance he joined the Royal Agricultural Society of England in 1860 and in 1866 was appointed agent to the Chillingham estates of the Earl of Tankerville. His abilities were soon recognised with his appointment to the directorship of the Royal Show, a post which he held from 1875 until 1892 and again in emergency in 1905.[vi]

Concerned to enhance the standing of the Royal Show, he addressed many of the sensitive important issues surrounding the credibility of such events. In 1871 he drew attention to the responsibility of judges in carrying out their task particularly with the newer breeds of stock. Great differences of opinion on standard type were a problem to the society. Jacob Wilson not only persuaded the society to give clearer guidance but introduced a condition that judges in their report should indicate the reasons for their decisions. vii

Such was his contribution to the premier national show that he became something of a national institution. At the conclusion of the Jubilee event at Windsor in 1889 Queen Victoria sent a command that Jacob attend the dinner party at Windsor Castle. There she expressed her satisfaction in a way very pleasing to the society, conferring a knighthood on Jacob. At a national level he played a major role in bringing forward the Contagious Diseases (Animals) Acts of 1878, 1884 and 1896. In 1888 he chaired a commission of enquiry into Bovine Pleuropneumonia. viii As an early beneficiary of a scientific education in agriculture he promoted the benefits and served as a patron of the Aspatria Agricultural College, second only to the Royal in its foundation. In 1884 in further honours came with the presentation of a purse of 3,000 guineas in recognition of his contribution to the industry. The subscriptions were limited to 20 guineas, indication of a list of some length and which was headed by HRH the Prince of Wales. ix In the birthday honours of 1905 the King made him a Knight Commander of the Royal Victoria Order (KCRVO). Sir Jacob undertook the directorship of the 1905 show following several difficult years. The effort took its toll and three days after the show he was dead.

Testing new ideas

Responding to change is about leadership and demonstration of the possible. Testing new ideas in the 19th century was in reality risky and expensive, leaving the task mainly to the larger farmers and estates. Within Eden are examples of exceptional improvers well acquainted with the most progressive ideas of the time and committed to test them in practice. Such experimentation allowed the wider farming population to assess the relevance of innovative practice to local farming systems.

John Grey of Dilston, the steward of the Greenwich Hospital estates, played a particular role in the Alston Moor area. The estate extended to over 34,000 acres of farmland with 290 tenancies.[x] Grey took up his post in 1833 soon after the enclosure of 20,000 acres of Alston Moor, much of which belonged to the Greenwich Hospital estates. Spending typically 5 to 8 hours a day on horseback, John Grey oversaw the regeneration of the holdings through buildings, drainage and walling. In addition to £8,000 a year rental from the lead mines the estate drew £25,000 from the farms, a figure which rose during Grey's stewardship to £40,000 in 1863. He also laid the foundations for later improvements in a community where lead mining was the chief interest and where land had served primarily as accommodation for miners.[xi]

Through investment which exceeded £100,000 and with the establishment of a farmers club and library, John Grey encouraged a positive and early response in an isolated community. The reputation of the area was such that Grey was a regular adviser to the government and on one occasion to the Commissioner to the Emperor of France.[xii]

Guano from South America

Such efforts to encourage adoption of progressive ideas reflected at least in part the self interest of landowners. The Lowther estate actively pursued such policies with Lord Lonsdale pioneering the introduction of improved shorthorns and tile drainage, two of the key elements of improvement. Of equal significance perhaps was the importation of guano — seabird droppings from South America — as early fertilisers in the mid 19th century. Shiploads were transported by rail from Whitehaven to the Penrith area and transported in horsedrawn wagons to farms on the estate. Charging the guano at cost provided a real incentive to raise levels of food production dramatically. A clear example of personal indulgence in demonstrating the art of the possible was undertaken at Wasdale, above Shap. Over 1,200 acres were drained and limed. Six foot walls were erected for stock control and the area opened for public agistement. Crayston Webster, a Westmorand land agent, wrote: " Numerous were the critics and foreboders of failure", yet the objective was achieved. No doubt the site adjacent to the A6 road ensured that it was subject to close scrutiny by locals and others travelling the turnpike.[xiii]

▲ *Sir Thomas Middleton, who helped establish Newton Rigg as a national leader in agricultural education.*

Other landowners lead in different ways. The Howard estate based on Greystoke extended into the Lake District to the west and south as far as Glencoyne and Deepdale in Patterdale. Henry Howard (1791-1856), a relative of John Christian Curwen, engaged in agricultural improvement and the establishment of the Penrith Agricultural Society. In the late 19th century Henry Charles Howard (Harry) played a major role in promoting public provision for agricultural education. In 1894 he chaired a meeting to establish a Northern Dairy School by the county councils of Northumberland, Cumberland and Westmorland. When Northumberland withdrew Harry Howard held the partnership of the twin counties together, negotiated a lease on Newton Rigg farm and selflessly proposed Frederick Punchard of Underley estates as the first chairman. [xiv] Harry Howard made a major contribution to the 1908 Reay Commission on agricultural education, concerned with identifying a model for farm institutes. He also promoted the idea of school gardens as a means of rural education. [xv]

Champion of many causes

He served as chairman of the Newton Rigg governors as did Lady Mabel Howard after the passing of her husband. Her work in the area was energetically and persuasively undertaken. She championed many causes including the Young Famers Clubs and the work of the Land Settlement Association. [xvi]

The guidance of such progressives depended ultimately on the adoption of ideas and practices by farmers and farmworkers. Perhaps the greatest attribute of agrarian change in Eden has been the responsiveness of the farming community many of whom have and had close associations with Penrith Agricultural Society.

Increasingly from the late 19th century science overtook the art of agriculture, at least in part. This was a change too far for the leading landowners. Increasingly experimental work was undertaken by publically funded institutions. Such work reflected the increasing confidence of the improvers in the motto "practice with science". In 1891 the Durham College of Science began to develop experimental programmes and demonstration sites to attract the interest of farmers. In 1894 Cumberland County Council secured an agreement to establish 20 experimental sites in the county. Subsequently Westmorland joined the scheme having run an independent programme for some years.

In 1896 Cockle Park was added to Armstrong College to provide facilities for research. The first three directors were men of vision and ability who formed important partnerships with the Cumbrian farming communities to mutual benefit. Sir William Somerville devised many classical experiments in Cumberland, leading to the development of the modern grass seeds mixtures and the use of basic slag as a fertiliser.

▲ *Douglas Gilchrist, a pioneer of modern grass seeds mixtures.*

▲ *Sir William Somerville, deviser of important agricultural experiments carried out in Cumberland and Westmorland.*

▲ *Franklin Englemann and Newton Rigg principal Jim Hall filming "The Other Man's Farm", one of the pioneering ventures using television for technical advisory purposes.*

▲ *Ike Dent, from Brampton, Appleby, walling with William Rawling, of Ennerdale and Ianthe Thistlethwaite, of Sedbergh, in about 1970.*

In 1908, at a parliamentary enquiry, Somerville referred to the experiments which he had established with W. T. Lawrence of Newton Rigg and in the North East as the longest in the experimental history of Northern England and exceeded nationally only by those at Rothamstead.

Sir Thomas Middleton followed zealously in the cause of education. His work with Newton Rigg played no small part in establishing the reputation of the institution as a national leader. Douglas Gilchrist followed and remained in the post for 25 years, a benefit which perhaps was only fully recognised with hindsight. Sir George Stapledon, the pre-eminent grassland researcher, paid tribute to Gilchrist in 1944 saying that "it was to Gilchrist that we should give the chief credit of having brought together the phosphate, the wild white clover, the sensible seeds mixture, the greatest needs of the farmer, all together in the ley".

Director of hill farming research

For three decades after the Second World War advice was provided free by a government anxious to raise levels of self sufficiency in food. One of the key scientific organisations to be established was the Hill Farming Research Organisation (HFRO) which is now integrated into the Macaulay Land Use Reseach Institute, undertaking fundamental research on an international basis. The first Director of the HFRO was Arthur Wannop OBE. from Blencowe, who brought leadership, direction and sound guidance. In the formative years Arthur Wannop established a sound understanding of farming systems. [xvii]The organisation and delivery of the service depended on partnership between researchers, advisers and farmers supported by institutions and experimental husbandry farms. Credit is rightly due to the men and women working with farmers to raise levels of production and efficiency.

At the end of the second millennium such reflections are but a memory, the circumstances of the industry constantly changing. Many individuals have given and continue to give their time, understanding and vision to the benefit of agriculture and its ancillary services. Many will remember the contribution to farming and the community of Johnny Friend Herdman, of Garrigill, for many years Chairman of governors at Newton Rigg. Joe Harris has established something of a family tradition in working with the Royal Agricultural Society, a role that his son John now undertakes.

▲ *King of the Fells — shepherd Bill Teasdale, from Caldbeck, the great fell runner of the 1950s and 60s.*

▲ *Herdwick festival at Wasdale, February, 2000, organised by the Fells and Dales EU Leader II Program. Third from left is Jean Wilson, Penfold Farm, Matterdale, chairman of the Herdwick Sheep Breeders' Association in millennium year*

▲ *Tom Fishwick, the Long Sleddale shepherd, with his dog Ben at the head of Mardale after driving stray sheep over Gatesgarth Pass for Mardale Shepherds' Meet. The meet was originally held at the Dun Bull Hotel which was "drowned" with the village when Haweswater reservoir was created.*

John Dunning, from Orton, is carrying responsibility for the rural community at the Northwest Development Agency. Farmers like Les Armstrong and Peter Allen have taken up the reins from men such as Joe Raine in their direct representation of the farming interest at national and european level. Farming communities are of course much more complex than simply farming businesses. The work of those committed to Women's Institutes and Young Farmers Clubs addresses a wide range of issues. In the area of sustainable rural communities the district could have no better champion than Voluntary Action Cumbria under the direction of Kate Braithwaite whose work is recognised nationally as innovative.

---

i The Penguin Dictionary of Quotations Middlesex 1971 p 371.

ii Lonsdale H, The Worthies of Cumberland, London 1867 pp 213,216.

iii Chance Some Notable Cumbrians pp 33,34.

iv Lonsdale op cit pp 264, 265.

v Beckett J V and Heath J E eds Derbyshire Tithe Files 1836-1850, Chesterfield 1995 p xiv

vi Wright P ed. The Standard Cyclopoedia of Modern Agriculture and Rural Economy, London circa 1912. pp 160,161.

vii Scott-Watson J A The History of the Royal Agricultural Society of England 1839-1939. London 1939, p 50.

viii DNB Concise Edition Oxford 1993, p 3251.

ix Rea G G. Sir Jacob Wilson KCVO, Journal of the Royal Agricultural Society of England Vol 66 1905 p 10.

x Butler J Memoir of John Grey of Dilston, p 162.

xi Ibid p 163.

xii Ibid pp 212-214.

xiii Webster C, The Farming of Westmorland, Journal of the Royal Agricultural Society of England 2nd Series Vol 4 1868 p 35.

xiv Humphries A. Seeds of Change Penrith 1996, p 46.

xv Cd 4206 Report of the Departmental Committee on Agricultural Education in England and Wales, 1908 p 261.

xvi Humphries AB op cit p 60.

xvii Science and Hill Farming, Edinburgh 1979 pp i, ii.

▲ *Sheepdog handler George Hutton, Setmabanning, Threlkeld, receives a trophy from Ullswater Dog Day President Muriel Viscountess Lowther at Patterdale.*

The role and contribution of women

Pretty efficient — a line of dairy workers pictured at Newton Rigg in the 1950s.

Workers and watchers — a fair division of labour?

IN EDEN THE TERM FAMILY FARMING is synonymous with farming and is perceived as the idealised model; the natural and most fundamental characteristic of agriculture in the area. The future of farming in the area is about conserving, developing and sustaining family units.[i] The alternative is seen as somehow inferior and less wholesome. Women are self evidently at the heart of family farming, socially and economically, and at the turn of the millennium were once again emerging from the shadows to make a special contribution to the process of adjustment to change.

As in earlier times their role is multi-functional. If diversification means undertaking several roles at once then that is nothing new to farming women well used to undertaking work on the farm alongside domestic responsibilities, frequently caring for older and younger members of the extended family and contributing to community life. In the past women have tended to be at least in part invisible in economic terms and, being unpaid, frequently excluded from census data.

The input of women has historically been particularly high on farms without hired labour. Such farms characterise much of Eden. The input of women in Eden to farm work tended to be regular rather than the seasonal pattern found over much of England.

A further factor to note is that under the customary tenure in Cumberland and Westmorland the inheritance of the holding lay with the tenant rather than at the will of the lord of the manor. This would also have tended to encourage a larger number of women farmers than in much of England.

When Penrith Agricultural Society was formed the numbers of farm servants nationally was falling dramatically. In Cumberland and Westmorland however, farm service remained important. At Queen Victoria's accession they formed over 40 per cent. of the labour force and by 1852 the figure was still over 25 per cent. in Cumberland.[ii] In part the system persisted due to a lack of cottages.

"Inadequate arrangements"

E. W. Hasell, of Dalemain, expressed the views of many that the inadequate sleeping arrangements in the older farm houses were a considerable factor in immoral behaviour and the high rate of illegitimacy. By the 1870s things were much improved. Female servants were able to negotiate higher wages and conditions of service. Isabella Deans, a local farm servant, stated that local girls would no longer clean out byres. One of their principal objections to outdoor work was the cost of clothing. Isabella claimed that this took a lot of her half yearly £6 bearing in mind that female rates had been about half of those paid to men.[iii] Nevertheless the underlying attributes possessed by women relevant to dairying were already widely appreciated.

▲ *Bella Jackson, The Hause Farm, Martindale, feeding stirks in about 1900.*

The gentle, rosy dairymaid

"A cow will give a dairymaid more milk than she will to a man. We think a man, during the milking season, has seldom any business near a cow, his great, rough, hard hands, and still harder heart, rendering him unfit for a good milker, while a gentle, rosy dairymaid, with her kind words, soft hands, and 'so so my good bossy' seated on a three-legged stool, will fetch out the milk till the froth runs over the pail."[v]

In the 1860s milk was a comparatively unimportant commodity, consumption nationally being little more than a pint a week and butter and cheese generally unaffordable. By 1900 milk consumption had doubled and significant increases had occurred for butter and cheese. In the late 19th century little butter in the Lake District hotels was local, the majority being imported from America, Holland and Normandy. Good butter was made locally; the problem was inconsistency reflecting the variable conditions in farm dairies.

By the late 19th century American dairying had moved to a factory based system with its advantages of hygiene and consistency.[v] In Eden the main product was butter. There was little interest in cheese due in part to the value of skim milk for calf rearing compared to whey which was fundamentally pig feed.

Cheese and butter making were also physically demanding and time consuming, as was domestic textile production. This was particularly true of cheese, causing resistance from women who already contributed to regular farm work and catered not only for their family but generally for the living-in servants. By the 1890s farm servants had three meals a day plus 11 o'clocks and 4 o'clocks [vi]; no wonder their enthusiasm for cheesemaking was somewhat muted. Contemporary debate on social inclusion is a reminder of past shortfalls in service provision. At the beginning of the 20th century Westmorland had the highest death rate from infections and accidents at childbirth.[vii]

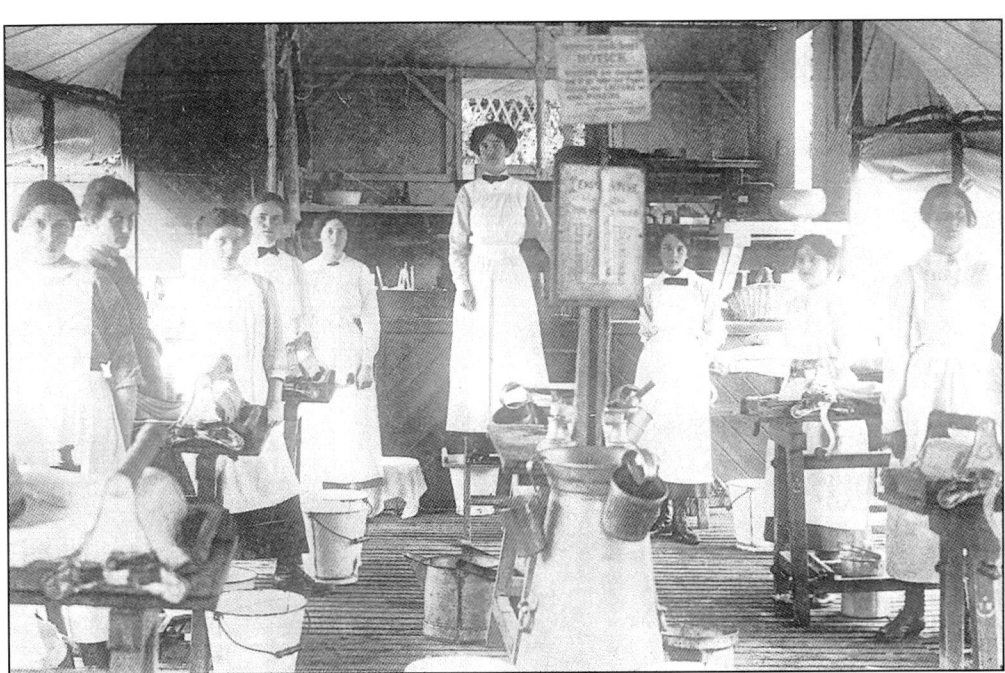

◀ *Cumberland County Council's Migratory Dairy School c1891. The school, in a marquee attached to a van, stimulated improvements in butter making. Penrith was the first community to request a visit.*

Butter making at Newton Rigg Farm School around 1900.

A crisis of depression

After 1875 farming fell into a crisis of depression which lasted more or less until the Second World War. The railways which connected Eden to the consumers in Manchester also opened up the New World and stimulated a cascade of cheap imports with which domestic producers struggled to compete. The one enterprise which successfully responded was that of dairying. Today we talk of added value, which is exactly the response which came from farming women in late 19th century Eden.

Dr Webb, the principal of the Aspatria Agricultural College, who had tried with some success to encourage women to attend classes, commented that women "are equal to if not superior in intellectual power to men; at any rate they are quicker to see the truth of a thing and act upon it". [viii]

The value of a woman possessing dairying skills was of a high order and enhanced matrimonial prospects. One Cumberland farmer giving evidence to a parliamentary enquiry admitted: "Aye she has been a gold mine to me, and if she hadn't been with me I don't think I could have done at all."

The potential to add value is well illustrated by the comparison of factory butter in 1887 which in Penrith was making 1s 4d per lb compared to 4d for local produce. In 1891 Cumberland County Council began the migratory dairy school using a horse drawn van, the first of its kind in England. The first booking was made for Penrith although the first classes were at Keswick. The Penrith classes were held adjacent to the Market Hall, from July 21st to 31st, 1891. Of course there was some resistance to change and the new scientific principles of separation and temperature control did not find universal acceptance. One Kendal farmer showed clear scepticism, exclaiming "dang ther dairy scheuls, my missus allus med good butter and got top price and I'll back her an t'ould way agin ther new notions fer out thou likes to lig doon". [ix]

Such setbacks were temporary and the improved prices were persuasive enough. Many of the students won awards at local shows and further afield. In 1902 Miss Hewitson, of Scales Hall, Skelton, won the buttermaking at the Royal Show at Carlisle for which she was awarded the Blamire medal. [x]

In the space of only 40 years the country was to experience the effects of two world wars. The "Great War" of 1914 - 1918 heralded a new aspect to conflict — the involvement of the whole population and the ability to use food blockades as a major weapon.

"War found us asleep"

As Charles Douglas wrote in the journal of the Highland Agricultural Society "war found us asleep". The cumulative effects of four decades of agricultural depression expressed themselves in land reversion, low fertility, collapsed drainage systems, few horses and fewer people. In that time butter imports had increased by 141 per cent., cheese by 180 per cent. and wheat by 238 per cent.

Strangely, the war was perceived as something of a continental affair that would by over by Christmas. Naval supremacy was still strong in the British psyche until in 1917 the U boats reduced food stocks to three weeks' supply. [xi]

In 1916 the War Agricultural Executive Committee had opened a register for women volunteers, some 250 registering from Cumberland and 300 from Westmorland. Training courses began in 1917 at Newton Rigg with additional practical facilities available at Brackenburgh and Greystoke Castle. In 1918 a rally and demonstration were held in Penrith. The parade was led by the S Lancashire regimental band followed by lorry displays. Lady Mabel Howard presided. [xii]

▲ Sunday cycling. Sara Jane Robson, of Serwborwens Farm, near Penrith, and two friends.

Extracts from a letter written by a World War I Westmorland trainee at Newton Rigg

IT WAS WHEN WE had so few patients at our VAD hospital that I first thought of taking up work on the land. In every newspaper there were appeals to women to help their country by working on farms and so not allow a decrease in food from ur own land.

On June 19th I went to Newton Rigg ... I took a special course with seven others arranged for those only taking up this work for the duration of the war. We arose at 5.30am every morning and if it was our day for milking it was advisable to waken earlier still and so be able to choose a good cow which wasn't in the habit of putting its foot in one's milk can or overturn the milker.

At 7.30 the breakfast was rung — and I can assure you that it was a welcome sound. Hoeing turnips filled up much of our time. Never before did I realise there was so much work and worry with them. In the early stages there is the turnip fly to be overcome ... then there is side hoeing or weeding, a very tedious job and one's back feels as though it could never be straightened again! When the sun shines one gets baked, for there is no shelter — and when it rains and the wind pierces through one it is dreadful, but the warm glow that follows the rain is perhaps worth the misery. We tried to mow with a scythe, but the end of it would either dig into the ground or miss half the grass. Every morning the war workers went into the dairy to make butter ... we each had a little churn and a worker and were highly delighted if we got a good grain and our butter wasn't sticky.

The thistles were growing fast in the pastures, so one dull morning we all turned out each with a different instrument. The grass was hanging with water, and soon heavy rain came on, but we didn't mind. Not till we were almost wet through did we go to shelter under a hedge where all the drippings ran down our necks.

The day before we worked hard in the eight acre hayfield from 9am until 9.30pm. We spread it, put it in rows, plaited and cocked it — I don't think I have ever worked so hard before. Then on turning back at the gates, before leaving, what a glorious feeling of satisfaction ... The field was finished, all but the carting ... We enjoyed this work better than the lectures on cattle and their different foods, the rearing of calves, the rotation of crops ...

Since my month's training I have given my services to a farmer at Crosthwaite, and helped with the hay, getting home each night about 9.30pm after a 40 minutes' cycle ride. They only had one man ... instead of the usual three, so the wife and I had to do all the forking on to the cart, then to the hay mow, where we mounted a ladder till we reached the beams. It was a very dark and dusty job ... This I think is the hardest of all, for you cannot stop, or you will have mountains of hay all around you.

But oh what an appetite ... Mine was even better here than at Newton Rigg. It was one of the weeks of brilliant sunshine, and the oatmeal and water that we took with us to the field was a true blessing. Tea was always welcome ... brought to us in the field in a big basket, by one of the children. I am going to help the same farmer with his harvest — they tell me it is much harder work than haymaking, but I shall have to wait and see.

Aimee Tomlinson.

(Source: Stramongate School Magazine vol 4 number 11 nd.)

▲ *Land Army girls armed with ferrets prepare for a rat hunt in a Caldbeck farm stackyard during the 1939-45 war years when it was reckoned that the nation's rat population exceeded that of the humans.*

◄ *A successful hunt. The ferrets have done their work and a line of dead rats is the result.*

▲ *Workforces on farms were supplemented during the Second World War by prisoners of war. Leading the horse drawing a hay turner at Moss Thorns Farm, near Penrith, is Rudi, a German prisoner at the Merrythought camp beside the Penrith-Carlisle A6 road. On the right is the farmer Charlies Fisher and on the machine two of his children, Dorothy and Jamie, and a young visitor.*

Penrith was host to Belgian refugees who benefited by the free supply of fruit and vegetables from the Newton Rigg gardens. Some 60 tonnes of sugar was distributed to fruit growers in the two counties for jam production. [xiii]

Only 71 workers were trained but the main benefit perhaps not anticipated was the ability to respond more quickly next time. There were a number of awards for service including 24 presentations for courage and endurance. One of these went to D. S. McCrae, a Cumberland girl and an outstanding individual who had the entire charge of a large firm of contracting engineers. She gained 100 per cent. in the tractor test and ploughed a field that soldier workers had refused to attempt. [xiv]

World War Two

The second war did not find us asleep. Winston Churchill gave an early and clear lead to farming.

"Today the farms of Britain are in the front line of freedom" (speech to the NFU Oct 14th, 1940)

During the conduct of the war self sufficiency in food rose from 30 per cent. to 60 per cent. Technical progress had been made since 1918 in fertiliser use, mechanisation and in the establishment of the Milk Marketing Board. These advances, though not fully exploited under the depressed economy of the thirties, were quickly and economically applied following the outbreak of war. During the war the number of tractors increased threefold and there was from the start a determination to reduce reliance on imports. The Land Army training began at Newton Rigg in September, 1939. The first course included 11 trainees from Cumberland, 7 from Westmorland and 14 from London. Subsequent groups included many volunteers from Tyneside.

The first volunteers to be trained included a number of local girls including:

Annie Stephenson (29 yrs) from Leadgate. Alston.
Dorothy Thompson (17 yrs) from Sandhill, Alston.
Annie Taylor (32 yrs) from Hill Crest, Crosby Ravensworth.
Hilda Irving, from Aughertree, Ireby. [xvi]

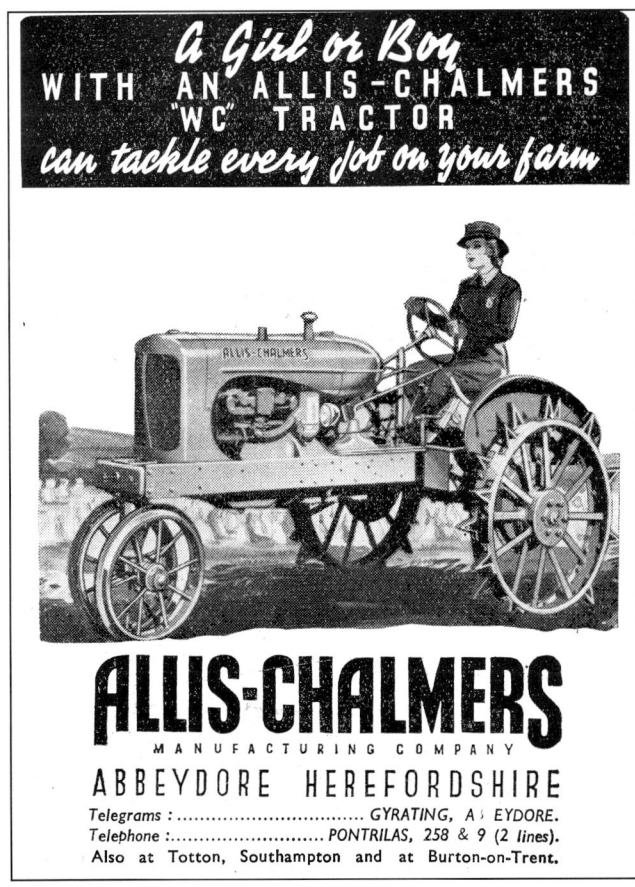

▲ *Land Army trainees in September, 1939, on the first course held at Newton Rigg Farm School.*

Naturally, farmers were cautious of the ability of enthusiastic but in some cases very inexperienced girls after a few weeks' training. Agricola, writing in the Westmorand Gazette in January, 1940, expressed doubts about the ability of women to manage tractors on Westmorland farms but conceded their suitability for dairy and poultry work. The courage and stamina of many of the volunteers is a matter for respect and admiration though the Women's Land Army, perhaps of all the women's volunteer groups, has been inadequately appreciated. During the war the WLA comprised 10 per cent. of the workforce. Local hostels included Roundthorn and Merrythought which provided accommodation primarily for those engaged in group work. Those employed as individuals generally lived in on the farm. Group working applied to such tasks as threshing, rat catching and fruit culture. In retrospect the courage and commitment of women often from urban situations to cope with changes in work, the vagaries of the Cumbrian climate, exposure to a different culture and real isolation deserves proper recognition.

Be gentle when you touch bread,
Let it not lie uncared for, unwanted,
Too often bread is taken for granted.
There is such beauty in bread,
Beauty of sun and soil,
Beauty of patient toil,
Wind and rain have caressed it,
Christ often blessed it.
Be gentle when you touch bread.

(verse in a Land Army Christmas card) xvi

To the crises which began in 1875, 1914 and 1939 may be added that which began in the last years of the second millennium. This is not the place to discuss in detail the causes and nature of the difficulties of the present time which bear on farming families and ancillary businesses.

In addition to the economic decline there has been a massive increase in regulation and legislation, adding an unprecedented burden of administration.

Whatever analysis future historians may make it is without doubt that women will be seen to have played a significant role in responsiveness. In almost every farming household they play a major role in administering records and paperwork. In many cases they are moving as quickly as practicable to longer term solutions utilising information technology.

As in previous difficult periods change brings pain to many but some opportunities. In looking for additional streams of income women are playing a critical role on and off the farm. It is possible and likely that history will judge that women will have played a role in the early 21st century at least comparable with their forebears in the crises which began in 1875, 1914 and 1939.

---

i O'Hara P. Partners in Production, Oxford 1998, p 1.
ii Humphries A. B. Agrarian Change in East Cumberland 1750 -1900 unpublished M Phil Thesis University of Lancaster 1993 pp 372,373.
iii Ibid p 370.
iv McMurry S. Transforming Rural Life, Baltimore, 1998 p 78.
v Ibid p 2
vi Humphries A. B. op cit. p 374
vii Horn P. Victorian Countrywomen, Oxford, 1991 p 221.
viii Humphries A. B. Seeds of Change, Penrith, 1996, p 12.
ix Ibid p 27.
x Oral information from Mrs D. Slater.
xi Sackville-West V. The Women's Land Army, London, 1944 p 9.
xii Humphries op cit p 46.
xiii 22nd Annual Report, Cumberland and Westmorland Farm School p 3.
xiv Journal of the Ministry of Agriculture Fisheries and Food, 1919 pp 754, 951.
xv MSS Register of Students.
xvi Sackville-West V. op. cit. p 40.